WORKSHOP PHYSICS® ACTIVITY GUIDE

THE CORE VOLUME WITH MODULE 1: MECHANICS I

Kinematics and Newtonian Dynamics
(Units 1-7)

PRISCILLA W. LAWS
DICKINSON COLLEGE

with contributing authors:
ROBERT J. BOYLE
PATRICK J. COONEY
KENNETH L. LAWS
JOHN W. LUETZELSCHWAB
DAVID R. SOKOLOFF
RONALD K. THORNTON

JOHN WILEY & SONS, INC.
New York Chichester Brisbane Toronto Singapore

Cover: Kinesthetic experiment to experience centripetal force.
The cover photo shows *Workshop Physics* students, Joshua Clapper and Jill Oppenheim, using *Kinesthesia 2*—a two-dimensional moving platform—for a centripetal force experiment. The platform is being pulled tangent to a circle by one student. As a result, the other student experiences a centripetal force on her hands and arms. This force is exerted by a rope attached to a post at the center of motion. This activity, which occurs in Unit 6, allows students to experience centripetal force directly and do a quantitative experiment to establish the relationships between F_c, v, and r.

Acquisitions Editor: Stuart Johnson
Senior Marketing Manager: Catherine Faduska
Assistant Marketing Manager: Ethan Goodman
Senior Production Editor: Katharine Rubin
Manufacturing Manager: Mark Cirillo
Photo Editor: Hilary Newman
Illustrations: Key West Graphics, Kim Banister, Erston Barnhart, and Noel Pixley
Cover Design: Michael Jung/Hans Pfister (photo)

This book was set in Aldus Roman and printed and bound by Courier-Westford.
The cover was printed by New England Book.

Recognizing the importance of preserving what has been written, it is a policy of John Wiley & Sons, Inc. to have books of enduring value published in the United States printed on acid-free paper, and we exert our best efforts to that end.

The paper in this book was manufactured by a mill whose forest management programs include sustained yield harvesting of its timberlands. Sustained yield harvesting principles ensure that the number of trees cut each year does not exceed the amount of new growth.

Library of Congress Cataloging in Publication Data:
Laws, Priscilla W.
Workshop physics activity guide/Priscilla W. Laws
p. cm.
Includes bibliographical references.
ISBN 0-471-10957-6 (pbk. : alk. paper)
1. Physics - Experiments. I. Title
QC33.L39 1995 95-37038
530'.078-dc20 CIP

Core Volume with Module 1 ISBN 0-471-15593-4
Module 2 ISBN 0-471-15594-2
Module 3 ISBN 0-471-15595-0
Module 4 ISBN 0-471-15596-9

Printed in the United States of America

10 9 8 7 6 5 4 3 2 1

PREFACE

The principle of science, the definition almost, is the following: The test of all knowledge is experiment. . . . But what is the source of knowledge? Where do the laws that are to be tested come from? Experiment, itself, helps to produce these laws, in the sense that it gives us hints. But also needed is imagination to create from these hints the great generalizations—to guess at the wonderful, simple, but very strange patterns beneath them all, and then to experiment to check again whether we have made the right guess.[1]

—Richard Feynman

This Activity Guide was developed as part of the Workshop Physics Project that began in 1987.[2] Although this Guide contains text material and experiments, it is neither a textbook nor a laboratory manual. It is a student workbook designed to serve as the foundation for a two-semester calculus-based introductory physics course sequence that is activity-centered. The Guide consists of 28 units that interweave text materials with activities that include prediction, qualitative observation, explanation, equation derivation, mathematical model building, quantitative experiment, and problem solving. Students use a powerful set of computer tools to record, display, and analyze data as well as to develop mathematical models of physical phenomena.

The Activity Guide represents a philosophical and pedagogical departure from traditional physics instruction. Students who study science in lecture-based courses are presented with definitions and theoretical principles. They are then asked to apply this knowledge to the solution of textbook problems and the completion of equation verification experiments. A major objective of Workshop Physics courses is to help students understand the basis of knowledge in physics as a subtle interplay between observations, experiments, definitions, mathematical description, and the construction of theories. Workshop Physics students spend an equivalent amount of time solving problems and doing equation verification experiments as those who study under the lecture method, but instead of spending time in lectures, they make predictions and observations, do guided derivations, and learn to use flexible computer tools to develop mathematical models of phenomena.

There are several reasons for emphasizing the processes of scientific investigation and the development of investigative skills. First, the majority of students enrolled in introductory physics courses at the high school and college levels do not have sufficient concrete experience with physical phenomena to comprehend theories and mathematical derivations presented in lectures. Second, the current body of physics knowledge is overwhelming and the traditional method of trying to cover too many topics can lead to rote memorization. We believe that the only viable strategy is to learn some topics thoroughly and to master methods for learning other topics independently. Finally, we have found, through many years of student evaluations, that most students prefer this active method of learning.

In this preface, we describe some important factors involved in the use of the Activity Guide and the Workshop approach. We suggest ways for using the Guide in various settings, outline the topics covered, and discuss the learning sequence that serves a model for the development of activities. We have also summarized computer requirements and the role that additional assignments such as homework problems, textbook readings, and projects can play in enhancing learning.

---

1. R.P. Feynman, The Feynman Lectures on Physics, Ch. 1 (Addison-Wesley, Reading, MA, 1964).
2. P. Laws, "Calculus-Based Physics Without Lectures," *Physics Today*, Vol. 44, No. 12 (Dec. 1991); "Workshop Physics—Learning Introductory Physics by Doing It," *Change*, 20-27 (July/Aug. 1991); and "Workshop Physics—Replacing Lectures with Real Experience," *Proceedings of the Conference on Computers in Physics Instruction*, Addison-Wesley, Reading, MA, 1989.

USING THE GUIDE IN VARIOUS INSTRUCTIONAL SETTINGS

This Activity Guide was originally designed to be used in relatively small classes in an instructional setting that combines laboratory and computer activities with discussion. Students are strongly encouraged to work in collaborative groups of 2, 3, or 4 depending on the nature of each activity. The activities in the Guide were tested and refined over an eight-year period at Dickinson College in a workshop environment where sections of up to 24 students met three times a week for 2-hour class periods. This workshop environment has been shown to work with larger groups. In the Studio Physics classes at Rensselaer Polytechnic Institute, sections of up to 50 students meet in 2-hour sessions in a laboratory setting to engage in activities enhanced by the use of integrated computer tools.[3]

Many institutions have successfully used some or all of the units in both calculus-based and algebra-based courses in universities, colleges, and high schools. Instructors at Gettysburg High School, the University of Oregon, and Moorhead State University in Minnesota have found little difficulty in creating algebra-based versions of various activities. An instructor in a small program at Lord Fairfax Community College in Virginia taught both algebra-based and calculus-based Workshop Physics in the same classroom to save himself a considerable amount of instruction time. To facilitate the use of the Activity Guide in algebra-based courses, the instructor materials include recommended reading assignments for each of the units for both Halliday, Resnick, and Walker (Wiley's calculus-based text) and Cutnell and Johnson (Wiley's algebra-based text).

However, colleagues at other institutions both large and small have found creative ways of using activities designed for this Guide in various instructional settings. For instance, activities set up to help students understand Newton's Third Law involving the collisions of force sensors attached to carts have been used in a 2-hour workshop session at Moorhead State College, in an hour-long tutorial session at the University of Maryland, in an interactive lecture demonstration at the University of Oregon, and in a weekly laboratory session at the University of North Carolina.

Thus, we believe this Guide can also be used in courses with large classes where the schedule still includes a mix of lecture, recitation, and laboratory sessions. With careful planning, each unit can be divided into activities conducted as interactive lecture demonstrations with class discussion, activities launched during hour-long tutorials designed to replace recitation sessions, activities done in the laboratory, and those assigned as homework. Each set of activities has been annotated in the Guide to identify which of them can be presented in an interactive lecture demonstration format. Possibilities for different ways of using the activities are discussed in more detail in the instructor materials.

TOPICS COVERED AND THE MODULAR FORMAT OF THE GUIDE

No activity-centered course can move at the pace of traditional lecture courses. We have chosen to give students an opportunity to master the topics that are treated rather than have a superficial exposure to many topics. We have retained topics that are most often covered in conventional courses: Newtonian mechanics, thermodynamics, and electricity and magnetism. Units in wave motion, geometric optics, physical optics, and atomic physics are under development. Interested instructors can contact the authors about the possibility of testing new units. Also, an excellent activity-centered unit on geometric optics is currently available from Dewey Dykstra (Boise State University).

Although the coverage of traditional topics has been reduced by about 25%, students also complete one or two special units that deal with contemporary topics not usually treated in conventional introductory physics courses. We have developed three special topic units for this purpose that are quite popular with students: Unit 15 on Oscillations, Determinism, and Chaos; Unit 25 on Electronics; and Unit 28 on Radioactivity and Radon Monitoring.

Many instructors have adapted the Workshop Physics program for their course by introducing the units in a different order than they appear in the Guide. Often instructors only have time to

3. J. Wilson, "The CUPLE Physics Studio," *Phys. Teach.* 32, 518 (1994).

cover some of the topics. For these reasons, this Activity Guide is being distributed in four different modules. The first module, consisting of Units 1-7, is included in a core volume. This core volume contains a table of contents and index for all 28 units along with appendices that contain reference materials. The core volume is bound, three-hole punched, and perforated so it can be used as a stand-alone volume or stored in a standard binder. The remaining modules are also three-hole punched, but they are unbound for inclusion in a binder. All four of the modules are available in separate parts or packaged as a complete set:

The Core Volume with Module 1: Mechanics I
Kinematics and Newtonian Dynamics (Units 1-7)
(ISBN 0-471-15593-4)

Module 2: Mechanics II
Momentum, Energy, Rotational and Harmonic Motion, and Chaos (Units 8-15)
(ISBN 0-471-15594-2)

Module 3: Heat, Temperature, and Nuclear Radiation
Thermodynamics, Kinetic Theory, Heat Engines, Nuclear Decay, and Radon Monitoring (Units 16-18 & 28)
(ISBN 0-471-15595-0)

Module 4: Electricity and Magnetism
Electrostatics, DC Circuits, Electronics, and Magnetism (Units 19-27)
(ISBN 0-471-15596-9)

The Complete Set: Core Volume with Modules 1-4 (Units 1-28)
(ISBN 0-471-10957-6)

Instructors should not plan to use units in Modules 2, 3, or 4 without covering substantial portions of Module 1 contained in the core volume and giving students access to the reference materials in that volume. Some of the units in Modules 2, 3, and 4 require treatment of topics in Module 2 on momentum and energy. The Workshop Physics Instructor Guide contains a detailed prerequisite scheme for each of the Units. This guide should be consulted before planning the syllabus for a sequence of the Workshop Physics course and deciding which modules should be purchased by students.

LEARNING SEQUENCES

In general, a four-part learning sequence is used for each topic. Students start by making predictions that require them to examine their preconceptions about the phenomenon being studied. Second, they reflect on their observations and refine their conceptions. Third, they develop definitions and derive equations based on theoretical considerations. Finally, they perform experiments intended to verify theoretical predictions and apply their understanding of the phenomenon to the solution of problems. This learning sequence is adapted from recommendations of cognitive psychologists such as David Kolb and physics educators such as Robert Karplus and Roger Osborne.

In designing activities we have attempted to take advantage of new insights into the nature of student learning difficulties based on physics education research. Many of these difficulties have been identified by master teachers and researchers such as Arnold Arons, Dewey Dykstra, Lillian McDermott, Jim Minstrell, and Eric Rogers. The classroom-based research conducted by contributing authors David Sokoloff and Ronald Thornton on conceptual learning has also been essential to the refinement of the activities.

Because we find that many students have severe conceptual difficulties in the study of Newton's laws, we developed a *New Mechanics* sequence with several key elements.[4]

1. Concepts in kinematics and dynamics are initially developed for one-dimensional horizontal motion with visible applied forces in the absence of significant friction;
2. Students are asked to make additional observations that lead them to invent invisible and passive forces such as friction forces, gravitational interaction forces, normal forces, and tension forces in order to maintain the viability of the Newtonian schema for predicting motions;
3. Next the study of kinematics and dynamics is extended to two-dimensional phenomena such as projectile motion, circular motion, and motion on an incline; and
4. Finally, students study collision forces, the Law of Conservation of Momentum, and center-of-mass concepts before dealing with energy concepts.

Our assessments of student learning indicate that, when students complete a sequence of Workshop Physics activities in supportive learning environments, they can achieve significant improvements in conceptual understanding in mechanics, thermodynamics, and electricity. We are currently extending our assessments to obtain data on the potential of activity-centered learning in the enhancement of proportional reasoning, problem solving, and mathematical literacy.

COMPUTER TOOLS

Computers are powerful learning tools when used flexibly for displaying, graphing, and analyzing data. They enhance the rate at which students learn through observation and experimentation with natural phenomena. Ideally, instructors will have a computer available for every two students in a class. However, with fewer computers, it is often possible for students to work in larger groups or participate in interactive computer demonstrations given by an instructor. Computer-based activities that can be presented in an interactive demonstration mode have been identified in each section of the Guide.

There are three different software packages and associated computer hardware that students doing activities should have access to throughout the courses. All the computer software and accessories are available for both Macintosh Computers operating under system 7 and PC computers operating under Windows 3.

1. Spreadsheets

Although we have worked with Microsoft Excel 4 and 5 spreadsheets, other spreadsheets such as Claris Works will also work and perform better than Excel on computers with limited memory and speed. Currently the most frequent use of the computer involves the entry of data directly into a spreadsheet for further analysis and graphing. Spreadsheet calculations are also used as a tool for numerical problem solving as an alternative to integration. In some cases spreadsheet calculations are used for mathematical modeling. Modeling is introduced early in the introductory courses as a way of fitting linear data and then exploring the behavior of objects undergoing constant accelerations. It is also used for the study of simple, damped, and driven harmonic motion as well as for the exploration of the behaviors of the chaotic physical pendulum studied in Unit 15. A set of Workshop Physics Excel Tools have been developed so that students can select icons placed on a custom tool bar for creating scatter plots from selected data and for performing linear or polynomial fits to the data. These tools are distributed with the instructor materials.

4. Priscilla Laws, "A New Order for Mechanics," prepared for the *Proceedings of the Conference on the Introductory Physics Course,* Rensselaer Polytechnic Institute, Troy, New York, May 20-23, 1993.

2. Computer-Based Laboratory Tools

These tools are used extensively for the collection, analysis, and real-time display of data. A computer-based laboratory system consists of a sensor plugged into a microcomputer via an electronic device known as an interface. Sensors used in this Guide include motion, force, rotary motion, temperature, voltage, and magnetic field. Data that are collected can be transferred easily to a spreadsheet for additional analysis. These tools are described in more detail in Appendix B and are available from PASCO scientific and Vernier Software.

3. VideoPoint Software

Because of the difficulties we encountered in designing activities for the observation of two-dimensional motion, we developed the VideoPoint software to allow students to analyze digitized video frames. By using a Macintosh or PC computer outfitted with a video capture board along with a video camera, VCR, or videodisk player, students can capture and store the information from a sequence of video frames for later analysis. Then they can use the VideoPoint Software to display the frames and determine the coordinates of locations of interest on the frames. These data can be transferred automatically to a spreadsheet for further analysis. The VideoPoint software and a collection of QuickTime digital movies are distributed by PASCO scientific.

Very limited use is made of computer simulations as we feel that giving students experience with real phenomena is of primary importance. However, we recommend several simulations for use with activities and homework assignments. These include *Electric Field Hockey* and *Graphs and Tracks* (Physics Academic Software), *Inertia Games* (Bob Morse, St. Alban's School), *Kinetic Gas Theory* (Drexel University), and *Coulomb* (in Physics Simulations II, Intellimation).

It is very important to give students access to spreadsheet, VideoPoint, and simulation software outside of class periods for the completion of activities and for homework assignments.

APPARATUS

Because the investment in computer tools is substantial, we have tried to use standard physics apparatus and inexpensive items that can be acquired locally for most of the activities. We recommend an investment in the new low-friction dynamics carts and aluminum tracks distributed by PASCO Scientific for a number of the mechanics activities.

In several instances where standard equipment was not available to support vital activities, we have designed or co-designed apparatus for PASCO Scientific.[5] Apparatus designed specifically for the Workshop Physics Project includes one- and two-dimensional Kinesthetic Carts,[6] a Force Sensor Adapter Bracket, a Mass and Torque Balance, a karate board testing rig, a Chaotic Physical Pendulum, a Mass Lifting Heat Engine, and Faraday apparatus.

Whenever feasible, the instructor materials include directions for constructing apparatus we have helped to design even when it is available commercially. We have also described how to acquire duck pin and bowling balls at modest cost, cut karate boards out of cheap pine, construct a manometer out of soda straws and Tygon tubing, and so on.

5. Many of these items are described in the 1996 PASCO Catalog.
6. H. Pfister and P. Laws, "Kinesthesia—1: Apparatus to Experience 1-D Motion," *Phys. Teach.* 33, 214 (1995).

ASSIGNMENTS AND TEXTBOOK READING

After completing a set of activities, students need to reinforce what they have learned by doing textbook readings and homework assignments. A complete set of recommended assignments is published in the instructor materials. Some of these assignments are similar to textbook problems. Others require students to engage in additional conceptual reasoning and write essays. Still other assignments require students to use spreadsheets or the VideoPoint software developed for use with the Activity Guide.

We suggest that students read about topics they have covered in the Activity Guide in an introductory physics textbook after completing a set of related activities. The instructor materials include recommended readings from the latest editions of Wiley's calculus-based text book (Halliday, Resnick, and Walker) and from Wiley's algebra-based text book (Cutnell and Johnson). These text readings play an essential part in supporting the Workshop Physics Activity Guide but they are not the central focus of the course. Thus, the text serves as a reference. It is not necessary to require the purchase of a textbook in courses where students have access to textbooks in a study center.

STUDENT PROJECTS

One of the most exciting assignments in the Workshop Physics courses at Dickinson has been the end-of-semester projects that student groups choose for themselves. Although these projects are laboratory-based, they are not done during class time. We have eliminated the third in-class examination each semester to give students more time for projects. The goal of the project assignment is to help students extend their understanding of a topic of special interest, acquire additional experience with the process of scientific investigation, and enhance their ability to write well-organized papers based on their research. Students have devised projects to analyze drag trials, the grand jeté in ballet, the effectiveness of automobile air bags, the efficiency of heat engines, and cooling processes. Biology majors have studied hummingbird flight, crayfish motility, bird song frequency spectra, and exponential growth processes. Budding engineers have designed and constructed circuits for event timing and for flushing toilets automatically. Chemistry majors have studied enthalpy and music majors have done spectral analyses of their instruments.

STAYING UP TO DATE

Attending Workshops: Pat Cooney, David Sokoloff, Ron Thornton, Mark Luetzelschwab, and Priscilla Laws have been offering workshops on various aspects of teaching Workshop Physics and/or related curricula on a regular basis. These include: (1) a two-week NSF Summer Seminar on Interactive Introductory Physics Teaching each June at Dickinson College; (2) one-day workshops every six months at the national meetings of the American Association of Physics Teachers; (3) three-day Chautauqua workshops once each year; (4) several three-day workshops for two-year college teachers each year; and (5) a one-day PTRA workshop for high school teachers every second summer. Look for notices in the *Physics Teacher,* the *TYC Newsletter,* the *AAPT Announcer,* and the Chautauqua course flyer for information about upcoming workshops.

Internet Access to Program and Document Files: We are constantly improving our Activity Guide Units and software tools, making it difficult to keep up to date. For this reason, we have created a Workshop Physics Project Home Page. We will place a list of upcoming workshops on this home page, along with student-generated QuickTime movies and other current information about our projects. From time to time we may list corrections, substitutions, or new assignments for various units in this Activity Guide. To log onto this Home Page you should type: "http://physics.dickinson.edu/." (If you have problems logging on, call the Workshop Physics Project Office at (717) 245-1845 during normal Eastern time working hours.)

ACKNOWLEDGMENTS

All of us who were involved with this project owe a debt of gratitude to the Physical Science Study Committee for its pioneering work in the revitalization of introductory physics courses. Two individuals whose approach to physics teaching became popular in the 1960s deserve special mention for their insights into student learning difficulties – Robert Karplus of UC Berkeley and Eric Rogers of Princeton University. In addition, the work of Arnold Arons and Lillian McDermott of the University of Washington have provided inspiration for this work.

During the past eight years many people have contributed to the development of the Workshop Physics Project and this Activity Guide. First and foremost are the group of contributing authors: Robert Boyle (Units 16-18), Patrick Cooney (Unit 15), Kenneth Laws (Unit 25), John Luetzelschwab (Units 6-13 and 22-24), David Sokoloff (Units 3-7, 14, 16, 17, 22-24), and Ronald Thornton (Units 3-7, 14, 16, 17).

The following colleagues who have taught sections of the Workshop Physics courses at Dickinson College have contributed their insights based on the wisdom of experience. They include: Robert Boyle, Mary Brown, John Luetzelschwab, Windsor Morgan, Hans Pfister, Guy Vandegrift and Neil Wolf. Hans Pfister deserves special mention for the design of kinesthetic apparatus for the Workshop Physics courses. In addition, several sabbatical visitors have helped in the development of activities including Mary Brown from Dothan College, Desmond Penny from Southern Utah State College, and V. S. Rao from Memorial University in St. John's, Newfoundland. David Jackson, who is working on a new Workshop Physical Science curriculum, has given us suggestions on the Unit 5 activities.

The activities could not have been tested and refined without the work of several student generations of equipment managers and summer interns. They are Jennifer Atkins, Christopher Boswell, Joshua Clapper, Catherine Crosby, Ryan Davis, David Diduk, Christopher Eckert, Amy Filbin, Jake Hopkins, Michelle Lang, Mark Luetzelschwab, Despina Papazisus, Alison Sherwin, and Jeremiah Williams. I am also grateful to the 50 or so student assistants and graders and the approximately 500 students who have survived the Workshop courses as we tested and retested various activities.

Several Dickinson physics majors and project associates have developed software or software tools that have been used in the program including Grant Braught, David Egolf, Mike King, Sean LaShell, Mark Luetzelschwab, Brock Miller, and Phillip Williams. Several individuals in the Tufts University Center for Science and Mathematics Teaching who have rewritten early versions of computer-based laboratory software have redesigned portions of their software to meet our needs including Stephen Beardslee, Lars Travers, and Ronald Thornton.

The insights of colleagues from other departments and institutions have tested workshop activities or developed pedagogical approaches that have been helpful in the refinement of this Activity Guide. These colleagues are Nancy Baxter-Hastings (Dickinson College Department of Mathematics), Gerald Hart and Roger Sipson (Moorhead State University), Robert Morse (St. Alban's School), E. F. Redish and Jeffrey Saul (University of Maryland), Mark Schneider (Grinnell College), Robert Teese (Muskingham College), Maxine Willis (Gettysburg High School), William Welch (Carroll College), and Jack Wilson (Rensselaer Polytechnic Institute).

Several administrators at Dickinson College have arranged for financial support for purchasing equipment, remodeling our classroom and equipment storage areas, and providing facilities for project staff. These individuals include President A. Lee Fritschler, Deans George Allan and Margaret Garrett, the treasurer, Michael Britton, and grants officer Christina Van Buskirk.

Individuals from the commercial sector have helped with the design, production, and distribution of hardware, software, and apparatus needed for the activities in this guide. They include: David and Christine Vernier of Vernier Software, Paul Stokstad and David Griffith of PASCO Scientific, Rudolph Graf of Science Source, and Ron and Wendy Budworth of Transpacific Computer Company.

Workshop Physics Project support staff who have helped with the production of this Activity Guide include Susan Greenbaum, Gail Oliver, Susan Rogers, Pam Rosborough, Virginia Trumbauer, and Maurinda Wingard. Kim Banister, Erston Barnhart, Kevin Laws, Virginia Laws, and Noel Pixley have helped with the artwork. Wiley editors Clifford Mills and Stuart Johnson with the help of Katharine Rubin have coordinated the Activity Guide production effort.

Major support for this work was provided by the Fund for Improvement of Post Secondary Education (Grant #G008642146 and #P116B90692-90) and the National Science Foundation (Grant #USE-9150589, #USE-9153725, #DUE-9451287 and #DUE-9455561). Project Officers who have provided administrative and moral support for this project are Rusty Garth, Brian Lekander, and Dora Marcus from FIPSE, and Ruth Howes, Kenneth Krane, and Duncan McBride from NSF.

Priscilla Laws
Department of Physics and Astronomy
Dickinson College
Carlisle, PA
July 1995

On behalf of contributing authors Robert Boyle, Patrick Cooney, Kenneth Laws, John Luetzelschwab, David Sokoloff, and Ronald Thornton

This project was supported, in part, by the Fund for Secondary Education and the National Science Foundation. Opinions expressed are those of the authors and not necessarily those of the foundations.

CONTENTS

Name ______________________ Section __________ Date __________

UNIT 1: INTRODUCTION AND COMPUTING

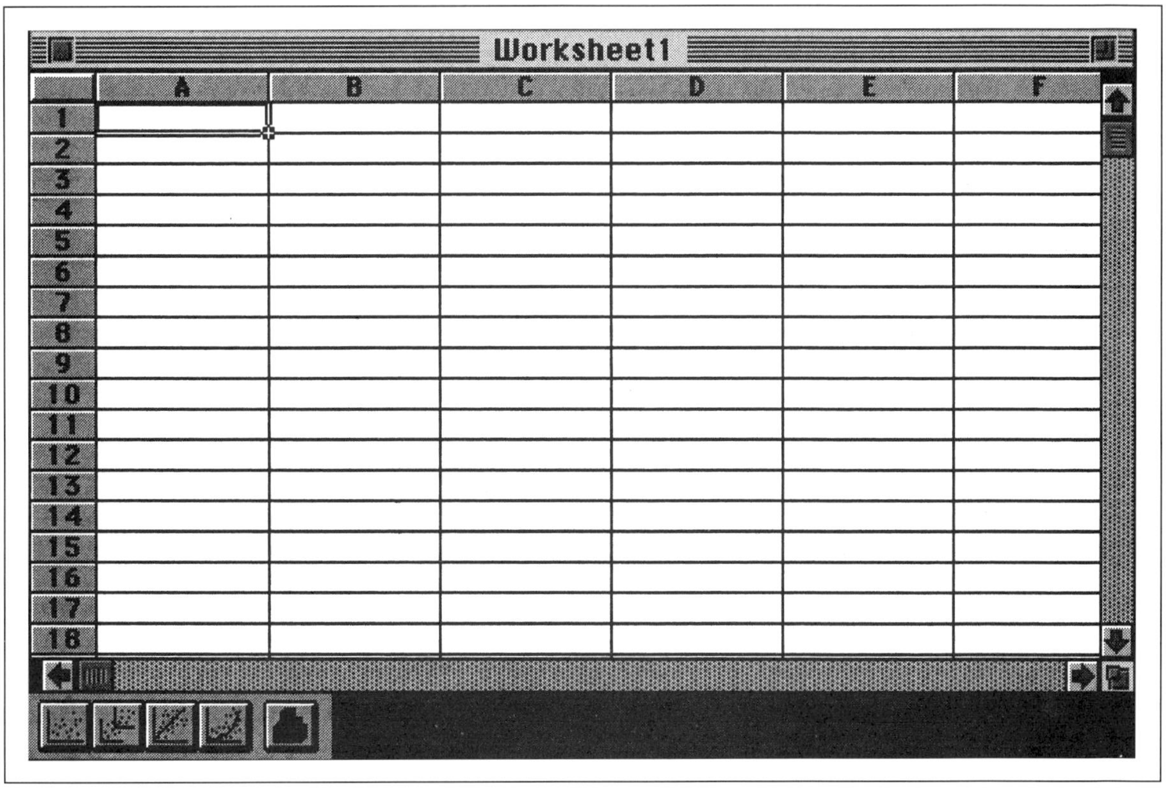

Since 1980 computer spreadsheets have revolutionized the field of accounting and business. By putting numbers and text-based information in little rectangular cells on a computer screen, accountants can enter directions for how to use the information in each of the cells to perform calculations. If bad information is placed in a cell, the user merely has to update the information and the computer will automatically recalculate everything based on the new information. Charts and graphs can be generated automatically from information on a spreadsheet.

How can this vital new accounting tool help you learn physics? When you complete this unit, you will learn about a few of the many ways in which spreadsheets can be used in physics research, and begin to acquire a set of valuable new skills that will aid you in your career as a student and later in life as you pursue careers in business, law, public service, medicine, science, or mathematics.

UNIT 1: INTRODUCTION AND COMPUTING

. . . I now refuse to teach . . . unless a computer is available in the classroom. And I insist that students have access to computers outside of the classroom. . . . First, the powers of the computer can remove a great deal of drudgery. . . . Secondly, the graphics capabilities of personal computers are wonderful.

John G. Kemeny, 1988
Coinventor of the BASIC Programming Language
Professor of Mathematics and Former President, Dartmouth College

OBJECTIVES

1. To understand *Workshop Physics* goals and procedures.

2. To explore the nature of horizontal motion in free flight.

3. To learn to load disks and access programs as well as to create, save, and print files with the type of computer used in this course.

4. To learn to use spreadsheet and graphing software to organize and perform calculations on data and then display it graphically.

1.1 OVERVIEW

Research and analysis are the basis of all knowledge in the sciences. Scientific research involves a constant interplay between several activities, including observing, reflecting, developing theories, and testing theories with experiment.

Some areas of research require tedious calculations or the use of mathematical equations that cannot be solved. In the past, scientists, being the lazy dogs that they were, tended to avoid mathematically difficult research. Since about 1960, the digital computer has changed dramatically both the areas of research in science and the way that research is performed. For example, a new field of physics, known as Chaos, has developed in the last few years. Chaos theory utilizes a new branch of fractal mathematics and the digital computer to find underlying patterns in the motion of systems that appear to move in a random and disorderly fashion.

Pioneering science teachers have brought the methods of learning science and doing science closer together. We are at the dawn of a new age in physics education, in which the use of the personal computer will become as commonplace as that of the pencil, electronic calculator, meter stick, and stopwatch.

From time to time, the computer will be used in this course to simulate a "real" event that is hard to observe directly. However, the primary function of the computer is not to teach you physics. Instead, the computer will serve as an active tool for the collection, organization, analysis, and graphing of scientific data. You will also use the computer to do mathematical modeling, which involves finding mathematical equations that best describe data. The computer is also useful for solving mathematical physics problems, performing theoretical calculations, and writing reports. The computer skills you learn should be invaluable to you in other courses and many other projects you embark on in the future.

DATA COLLECTION AND SPREADSHEET USE

1.2 HOW FAST DO YOU PITCH?

In this section you will be asked to predict and then measure your pitching speed. This section is designed to give you experience with making predictions and then taking measurements to test the predictions. You will be using both hand calculations and spreadsheet calculations to find speeds based on measurements of times and distances.

Suppose you were to go outside, stand at a comfortable distance from a classmate, and pitch a ball to him or her. How fast do you think you can pitch? **Note:** The distance from a pitcher's mound to home plate is 60.5 feet and a world-class professional pitcher can throw a baseball at a speed of just over 100 miles/hour.

Often, before we make observations or take measurements for scientific purposes, we try to predict the outcome. Sometimes there's no particular basis for a prediction other than gut feeling; for the purposes of this course, we'll define this type of prediction as a *guess.* On the other hand, if you are trying to develop a scientific explanation or theory for a phenomenon that has not yet been tested experimentally, scientists would define your prediction as a *hypothesis.** Many times a prediction is somewhere between a guess and a hypothesis, in the sense that someone has done some reasoning and developed an explanation, but the explanation is not part of a formal scientific theory.

1.2.1 Activity: Predicting Your Pitching Speed

a. What's the fastest you think you can pitch comfortably (in miles per hour)?

b. If you have one or more reasons for your prediction, explain it.

c. Would you call your prediction a guess, a hypothesis, or something in between? Explain why.

* An alternative way to distinguish between a guess and a hypothesis is to endow only scientists with the right to develop hypotheses. This approach is taken by Milton Rothman in his book on the *Experimental Basis of Physics* (Dover, New York, 1989) who states: "If the guesser is a scientist and he has the intention of testing these guesses with a suitable set of observations, then he dignifies each guess by calling it a *hypothesis.*"

Fig. 1.1.

1.3 COLLECTING DATA ON PITCHING SPEEDS

To measure your pitching speed each group will need:

- 1 tennis ball (or baseball)
- 1 digital stopwatch
- 1 tape measure, 15 m
- 1 mitt, optional

Recommended group size:	3	Interactive demo OK?:	N

In this activity, you should use a pre-measured series of distances, and work with two partners. Each of you should pitch three times at a comfortable distance. Your partners can catch and time the flight of the ball. "Pros" might want to try pitching from a full 20-meter distance, which is longer than the standard major league pitching distance. *Please warm up a bit. Don't kill your arm!*

1.3.1 Activity: Pitching Speed Data

a. Fill in the data table in the space below for yourself and three other classmates and calculate the average time and speed for each person to two significant figures.

Name	Distance (m)	t_1 (s)	t_2 (s)	t_3 (s)	Avg. t (s)	Avg. speed (m/s)

b. Calculate the average speed in miles per hour you measured for your own pitch (using conversion factors found in any standard introductory physics textbook). Show all the steps in your calculation. How good was your prediction?!

c. When you are finished, record your name, your throwing distance, and the average time-of-flight for your pitches in a table on the classroom board and then copy all of the data into your class pitching summary table in Activity 1.5.1.

1.4 CALCULATIONS WITH COMPUTERS AND SPREADSHEETS

A computer spreadsheet is an electronic version of the large sheets accountants use to record and calculate things with columns of numbers. By entering formulas into the computer, a spreadsheet program can perform calculations for you. If you make a mistake and must replace an incorrect number, the spreadsheet program will recalculate everything automatically. Physicists often display and perform calculations on spreadsheets using data tables like the one in Activity 1.2 above. Thus, spreadsheets can be invaluable tools for data analysis.

In this activity, you will:

1. Load a spreadsheet program into your computer's memory;
2. Enter pitcher names and pitching data in the Activity 1.2.1 data table into the spreadsheet;
3. Calculate the average times and speeds of the baseball by entering formulas into the spreadsheet; and
4. Save your spreadsheet in a computer file so it can be reloaded later.

Instructions on how to do these tasks are included in Appendix A. **Note:** If your spreadsheet cells contain too many significant figures, you should use the "format" feature to reduce the numbers to the appropriate number of significant figures.

1.4.1 Activity: Pitching Speed Spreadsheet

How do the numbers in the spreadsheet (which you are creating by using your own pitching data) and the hand calculations in Activity 1.3.1 (using the same data) compare to each other? If they differ in any way, try to explain why.

1.5 AVERAGE TIMES FOR THE WHOLE CLASS

In order to practice using spreadsheets you can use the all-class pitching data in a homework exercise. Thus, you should copy the class data table in the space below.

1.5.1 Activity: Class Pitching Data Summary

	Name	Avg. t (s)	Distance (m)	Avg. speed (m/s)
1				
2				
3				
4				
5				
6				

	Name	Avg. t (s)	Distance (m)	Avg. speed (m/s)
7				
8				
9				
10				
11				
12				
13				
14				
15				
16				
17				
18				
19				
20				
21				
22				
23				
24				

MEASURING AND GRAPHING HORIZONTAL MOTION

1.6 MEASURING THE MOTION OF A BOWLING BALL

Fig. 1.2.

A key to understanding how to describe motion near the surface of the earth is to observe horizontal motions and vertical motions separately. Eventually, situations in which an object undergoes both horizontal and vertical motion can be analyzed and understood as a combination of these two kinds of basic motions.

Let's start with horizontal motion. How do you think the horizontal position of a bowling ball changes over time as it rolls along on a smooth surface? For example, suppose you were to roll the ball a distance of 6.0 meters on a fairly level smooth floor. Do you expect that the ball would move at a steady speed, speed up, or slow down? To observe the horizontal motion of a "bowling ball" you can use a duck pin ball which is slightly smaller but quite similar to a regulation bowling ball. You should have a smooth level surface of about 7 meters in length. Each group will need:

- 1 duck pin ball (or bowling ball)
- 3 digital stopwatches
- 1 tape measure, 10 m
- 1 roll of masking tape (for marking distances)

Recommended group size:	4	Interactive demo OK?:	Y

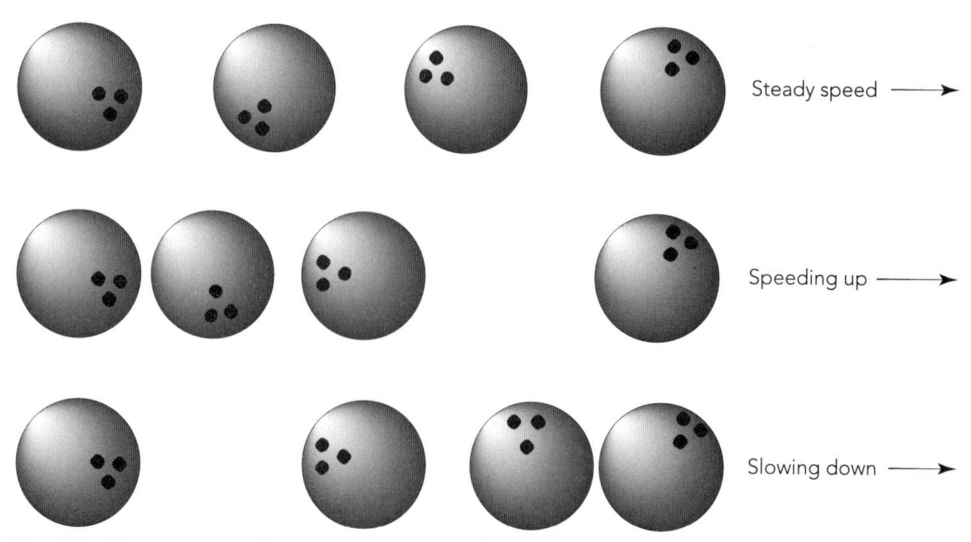

Fig. 1.3. Some guesses for the appearance of snapshots of a bowling ball taken at equal time intervals after it has been released by a bowler on a smooth surface.

1.6.1 Activity: Horizontal Motion

a. What do you predict will happen to the distance the ball moves as a function of time? Will the ball move at a steady speed, speed up, or slow down after it leaves the bowler's hand? Why?

b. Find a 7-meter length of smooth floor and mark off a starting line and distances of 2.00 m, 4.00 m, and 6.00 m from the starting line. Then

1. Bowl the ball along the surface.
2. Measure the time it takes to travel 2.0 m, 4.0 m, and 6.0 m.
3. Record the results below.
4. Open up a spreadsheet and transfer this data to it.

Note: This is a cooperative project. You will need a bowler and three people to time this. Practice several times before recording data in the following table.

Avg. t (s)	Distance (m)

c. Calculate the average speed, v, in m/s of the bowling ball as it travels the 6.00-meter distance.

1.7 GRAPHING THE HORIZONTAL MOTION

In this activity, graph your data for the distance your bowling ball traveled as a function of the rolling-time of the ball. This graphing should be done both by hand and on the computer. To do the computer graphing and print the graph, you can refer to the graphing instructions found in Appendix A and the instructions below.

1.7.1 Activity: Graphing Distance vs. Time

a. Fill in the units and the scale numbers in the graph below and plot the data you collected. You should plot a "fourth" data point on your graph by reasoning out what rolling-time you would measure for your ball if you were to roll it a distance of zero meters.

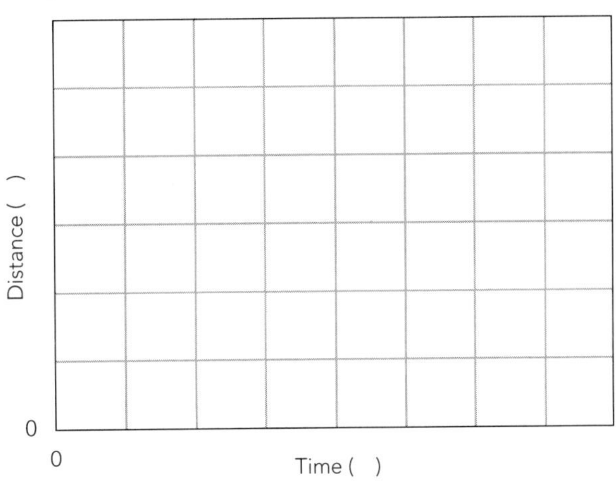

b. Following the directions in Appendix A, use the computer to create the same graph on the computer screen. Print out a copy of the graph. Affix a printout of the computer graph below. (Include a hand sketch if there is no printer available.)

1.8 HOW DOES THE BALL'S DISTANCE VARY WITH TIME?

We are interested in the mathematical nature of the relationship between distance and time for rolling on a level surface. Some definitions of mathematical relationships are shown in the sketches below.

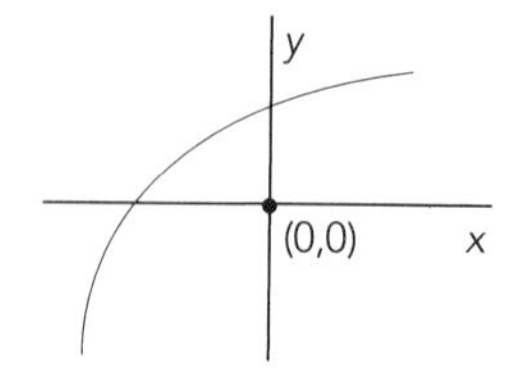

Fig. 1.4. y increases as x increases.

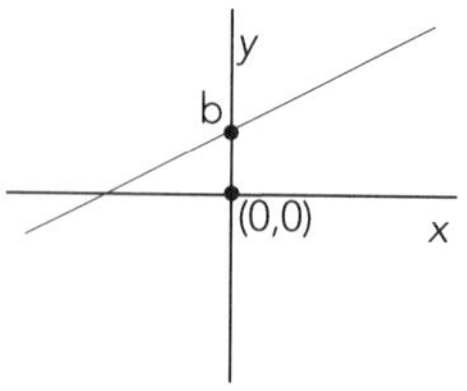

Fig. 1.5. y is a linear function of x that increases with x. The equation $y=mx+b$ can be used to describe the graph where m is the slope and b is the y-intercept of the line.

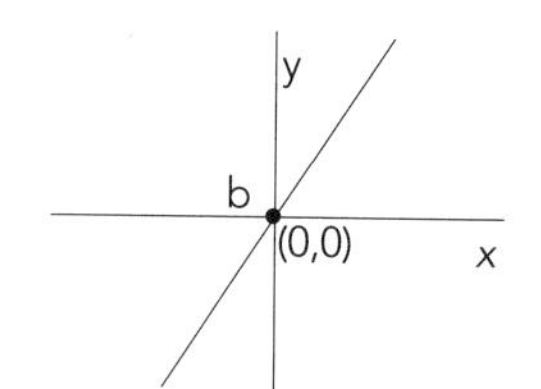

Fig. 1.6. y is proportional to x because the y-intercept, b, is 0 and the equation $y=mx$ can be used to describe the graph.

1.8.1 Activity: The Mathematical Relationships

a. By comparing the shape of the graph you have just produced with the sketches shown above, would you say that the distance increases with time, t? Decreases with time? Is it a linear function of t? Is it proportional to t? Explain.

b. How do the results compare with the prediction you made in Activity 1.7.1? Were you surprised?

c. What do you think would happen to the slope, m, of the graph, if the ball had been rolled faster? Would it increase? Decrease? Stay the same?

Name ______________________ Section __________ Date __________

UNIT 2: MEASUREMENT AND UNCERTAINTY

Physics is an experimental science and essentially all the mathematical theories developed by physicists are based ultimately on observations and measurements. It is possible for two investigators to measure a fundamental quantity such as the speed of light and come up with two different *numbers. Although neither investigator has made any measurement errors, they could decide that their results are the* same. *How is this agreement possible? It is because both investigators agree that the sources of uncertainty in their experimental measurements are similar in nature to the uncertainty in how a drunkard walks.*

When you complete this unit you should be able to explain why two scientists with different measurements can agree that they are actually the same and how an understanding of drunken movement can be used to resolve this paradox.

Louis Wolheim in a scene from the Academy Award-winning film *All Quiet on the Western Front.*

UNIT 2: MEASUREMENT AND UNCERTAINTY

THE
NORMAL
LAW OF ERROR
STANDS OUT IN THE
EXPERIENCE OF HUMANS
AS ONE OF MOST POWERFUL
GENERALIZATIONS OF NATURAL
PHILOSOPHY ◊ IT SERVES AS THE
GUIDING INSTRUMENT IN RESEARCHES
IN THE PHYSICAL AND SOCIAL SCIENCES AND
IN MEDICINE, AGRICULTURE AND ENGINEERING ◊
IT IS AN INDISPENSABLE TOOL FOR THE ANALYSIS AND THE
INTERPRETATION OF THE DATA OBTAINED BY OBSERVATION &
EXPERIMENT

Adapted from James Gleick's
Chaos: Making a New Science, 1988

OBJECTIVES

1. To define fundamental measurements for the description of motion and to develop some techniques for making indirect measurements using them.

2. To learn how to quantify and minimize sources of random uncertainty so that the precision of measurements can be enhanced.

3. To learn how to compensate for systematic error in measurements so that accuracy can be improved.

4. To explore the mathematical meaning of the standard deviation and standard error associated with a set of measurements.

2.1. OVERVIEW

Fig. 2.1.

At the beginning of this course you will focus on the task of developing a mathematical description of the motion of objects. The study of how objects move is known as *kinematics* and it can be conducted using only two fundamental types of measurements—*length* and *time*. For instance, if you are interested in determining how fast a pitched baseball is moving in a horizontal direction, you need to: 1) *define* horizontal speed (i.e., the meaning of "how fast") in terms of distance moved in space and time-of-flight; 2) *measure* the distance and time-of-flight of a moving baseball; and 3) *calculate* the speed of the baseball from your measurements. (Although physicists and philosophers can spend countless hours discussing concepts of space and time, in this course we will assume you have a sense of what they are without formal definition.)

The measurement of the speed of a pitched baseball is an indirect measurement. Almost any quantity has to be measured indirectly under certain circumstances. In this unit, you will devise methods for making direct and indirect measurements of distance. This should help you answer an age-old question: Is it possible to make exact measurements?

You will also make repeated measurements of the time it takes a ball to fall. Many sources of variation of the time–interval data will be explored including mistakes, systematic error, and random uncertainty. This timing activity will enable you to study statistical methods for determining the precision of measurements by quantifying the error and uncertainty associated with a set of repeated measurements subject to random variation. Finally, the mathematics of the Gaussian distribution used to describe your time measurements will be applied to a description of counting rates from a radioactive source.

The major goal of this unit is to help you determine, for measurements you will be making in this course, whether or not the results of a given experiment are compatible with theory.

DIRECT AND INDIRECT MEASUREMENTS

2.2. MEASURING LENGTHS DIRECTLY

We are interested in determining the number of *significant figures* in length measurements you might make. How is the number of significant figures determined? Suppose God could tell us that the "true" or "actual" or "real" width of a certain car key in centimeters was:

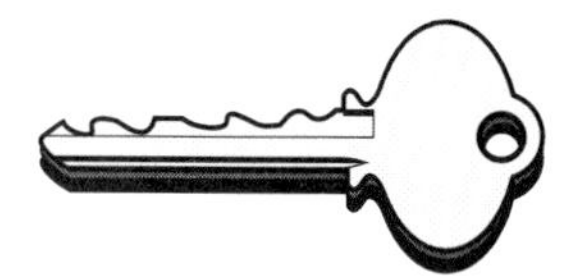

Fig. 2.2.

2.435789345646754456540123544332975774281245623 . . . etc.
(Sorry, but She got tired of announcing digits!)

If you were to measure the key width with a ruler that is lying around the lab, the precision of your measurement would be limited by the fact that the ruler only has lines marked on it every 0.1 cm. If the end of the key seems to be located between two ruled lines, you might be able to estimate to the nearest 0.01 centimeter how far the key edge is from the last mark. Thus, you might report that the best estimate for the width of the key is 2.44 ± 0.01 cm. This means you have estimated the key width to three significant figures.

If God announces that a pair of sunglasses is 13.276554577876542677 . . . cm wide, then upon direct measurement we might estimate their width to be 13.28 or 13.27 or 13.26 cm. In this case the estimated width is four significant figures and should report a measured value of $13.2\underline{7} \pm 0.01$ cm. Obviously, there is uncertainty about the "true" value of the right-most digit (which has been underlined for emphasis!).

Fig. 2.3.

Usually the number of significant figures in a measurement is given by the number of digits from the most certain digit on the left of the number up to and including the first uncertain digit on the right. In reporting a number, all digits except the significant digits should be dropped. The world is cluttered with meaningless uncertain digits. Help stamp them out! **Note:** If you have not encountered the idea of significant digits before, you can look up references to this concept in a physics text.

Fig. 2.4.

Let's do some length measurements to find out what factors might influence the number of significant figures in a measurement. You will need:

- 1 meter stick or tape measure

Recommended group size:	2	Interactive demo OK?:	N

2.2.1. Activity: Length Measurements

a. What factors might make a determination of the "true" length of an object measured with *your* meter stick or tape measure uncertain?

b. Find a door or window near you and measure its height with your measuring device several times. Create a table in the space below to list the measurements. Don't forget to include *units*—nag, nag!

c. In general, when a series of measurements are made, the *best estimate* is the *average* of those measurements. (See Appendix C for more detail.) In the space below list the minimum measurement, the maximum measurement, and the *best estimate* for the height you measured. Include your units!

$$h\,(\text{min}) =$$
$$h\,(\text{max}) =$$
$$h\,(\text{best est.}) =$$

d. How many significant figures should you report in your best estimate? Why?

e. For your object, what limits the number of significant figures most—variation in the actual height of the object or limitations in the accuracy of your measuring device or techniques? How do you know?

It's clearly impossible to make direct distance measurements without some uncertainty. To do so, you would have to have a measuring device with an infinite number of lines ruled on it with each line being an impossibly short distance from its neighbors!

2.3. ANNOUNCING A DISTANCE MEASURING CONTEST

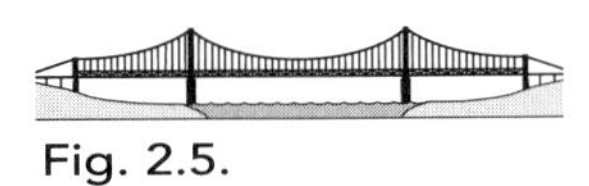

Fig. 2.5.

Direct distance measurements are as straight-forward as measurements can get. When it's not possible to make a direct measurement, some ingenuity is needed. Suppose you need to know the height of an object on campus such as a flagpole tip or an interesting feature on a building. How can you determine the vertical distance from the ground to point of interest in meters?

You might be assigned a vertical distance to measure. Prizes will be awarded for the "best" measurement as specified in the Contest Rules.

Recommended group size:	2	Interactive demo OK?:	N

CONTEST RULES

1. Students must work in pairs and each pair of students must hand in a write up describing the measurement technique and results and submit it jointly.

2. Only the following equipment can be used for the measurements:

   - 1 roll of Scotch tape
   - 1 tape measure, 15 m
   - 1 digital stopwatch
   - 1 ball
   - 2 meter sticks
   - 1 protractor

3. Measurement write-ups must be received by the deadline date and time announced by the instructor. Each write-up should be placed on an 8 1/2″ by 11″ sheet of paper and include:

   - the name of both cooperating partners
   - a description of what distance was being measured with a sketch of the buildings, flagpole, etc.
   - a lucid description of the technique used w/ any necessary equations and diagrams
   - a summary of data and calculations
   - the best estimate of the measurement (usually the average of all the really "good" measurements)
   - a discussion of the major sources of uncertainty in the measurement procedures.

4. Prizes will be awarded in each of the following categories with no individual being eligible for more than one prize:

   a. Smallest % discrepancy as determined by comparing your indirectly measured result with an accepted result determined by the course instructor(s). The following equation will be used in making the comparison:

   $$\%\ \text{discrepancy} = \frac{|\text{accepted distance} - \text{measured distance}|}{\text{accepted distance}} \times 100$$

   where the | | symbols represent absolute value.

   b. Most original method for a measurement that is within at least 10% of the accepted value.

2.4. STATISTICS–THE INEVITABILITY OF UNCERTAINTY

In common terminology there are three kinds of "errors": (1) mistakes or human errors, (2) systematic errors due to measurement or equipment problems and (3) inherent uncertainties.

2.4.1. Activity: Types of "Errors"

a. Give an example of how a person could make a "mistake" or "human error" while taking a length measurement.

b. Give an example of how a systematic error could occur because of the condition of the ruler when a set of length measurements are being made.

c. What might cause inherent uncertainties in a length measurement?

It is commonly believed that both mistakes and systematic errors can be eliminated completely if the person taking data is extremely careful and uses the best measuring equipment. However, inherent uncertainties do not result from mistakes or errors. Instead, they can be attributed in part to the impossibility of building measuring equipment that is precise to an infinite number of significant figures. The ruler provides us with an example of this. It can be made better and better, but it always has an ultimate limit of precision.

Fig. 2.6.

Another cause of inherent uncertainties is the large number of random variations affecting any phenomenon being studied. For instance, if you repeatedly drop a ball from the level of the lab table and measure the time of each fall, the measurements will most probably not all be the same. Even if the stopwatch was started and stopped electronically so as to be as precise as possible, there would be small fluctuations in the flow of currents through the circuits as a result of random thermal motion of atoms and molecules that make up the wires and circuit elements. This could change the stopwatch reading from measurement to measurement. The sweaty palm of the experimenter could cause the ball to stick to the hand for an extra fraction of a second, slight air currents in the room could change the ball's time of fall, vibrations could cause the floor to oscillate up and down an imperceptible distance, and so on.

2.5. REPEATED TIME-OF-FALL DATA

You and your partner can take repeated data on the time of fall of a baseball and eventually share it with the rest of the class. In this way, the class can amass a lot of data and study how it varies from some average value for the time-of-fall. For this activity you will need:

- 1 ball
- 1 digital stopwatch
- 1 meter stick, 2 m

Recommended group size:	3	Interactive demo OK?:	N

2.5.1. Activity: Timing a Falling Ball

a. Drop the ball so it falls through a height of 2.0 m at least 20 times in rapid succession and measure the time-of-fall to two significant figures. Be as exact as possible about the height from which you drop the ball, because your group and the rest of the class will use this data in a later activity. Record the data in the table below and enter it into a computer spreadsheet.

No.	t(s)
1	
2	
3	
4	
5	
6	
7	
8	
9	
10	

No.	t(s)
11	
12	
13	
14	
15	
16	
17	
18	
19	
20	

b. Use a spreadsheet to determine the average time-of-fall, $\langle t \rangle$, for your 20 measurements. Report the average value in the space below using *three significant figures.* Note: Be sure to save your spreadsheet since you will be using it again.

2.6. THE STANDARD DEVIATION AS A MEASURE OF UNCERTAINTY

How certain are we that the average fall-time determined in the previous activity is accurate? The average of a number of measurements does not tell the whole story. If all the times you measured were the same, the average would seem to be very precise. If each of the measurements varied from the others by a large amount, we would be less certain of the meaning of the average time. We need criteria for determining the certainty of our data. Statisticians often use a quantity called the standard deviation as a measure of the level of uncertainty in data. In fact, almost all scientific and statistical calculators and spreadsheets have a standard deviation function. The standard deviation is usually represented by the Greek letter σ (sigma; since sigma sometimes has other meanings in physics, we will designate the standard deviation by using a subscript: σ_{sd}). σ_{sd} has a formal mathematical definition that is described in Appendix C. The value of σ_{sd} is often used to measure the level of uncertainty in data.

In the next activity you will use the spreadsheet to calculate the value of the standard deviation for the repeated fall-time data you just obtained and explore how the standard deviation is related to variation in your data. In particular, you will try to answer this question: What percentage of your data lies within one standard deviation of the average you calculated?

2.6.1. Activity: Standard Deviation

a. Open the spreadsheet containing the time-of-fall data you collected in Activity 2.5.1. Calculate the standard deviation of the set of 20 measurements. (See Appendix C for instructions on how to use spreadsheet functions to calculate quantities such as σ_{sd}.) Write the calculated value σ_{sd} *with units* in the space below *using three significant figures.*

$$\sigma_{sd} =$$

b. Refer to the average you reported in Activity 2.5.1b and calculate the average plus the standard deviation and the average minus the standard deviation. Again report three significant figures and units.

$$\langle t \rangle + \sigma_{sd} = \qquad\qquad \langle t \rangle - \sigma_{sd} =$$

c. Use the *sort* command in your spreadsheet and determine the number of your data points that lie within $\pm\sigma_{sd}$ of the average you reported in Activity 2.5.1 b. Write the number of data points in the space below and calculate the percentage of data points lying within a standard deviation of the average.

d. Combine your results with those obtained by the rest of the class and then copy these results into the table. Once again, use *three significant figures.* Calculate the average time and the average % of data points that lie between $t - \sigma_{sd}$ and $t + \sigma_{sd}$.

No.	Investigators	$\langle t \rangle$ (s)	$\sigma_{sd}(\sigma)$	% Data $\pm\ \sigma_{sd}$
1				
2				
3				
4				
5				
6				
7				
8				
9				
10				
11				
12				
		Average time		Average % Data

e. Study the last column, which represents the percentage of data points lying within one standard deviation of the average. What does the standard deviation, σ_{sd}, tell you about the approximate probability that another measurement will lie within $\pm\sigma_{sd}$ of the average?

RANDOM AND SYSTEMATIC VARIATION

2.7. THE TIME-OF-FALL FREQUENCY DISTRIBUTION

How could you quantify the variation of your ball's time-of-fall data? You might characterize it by using the standard deviation, as you did in the last session. Another approach is to plot a type of graph known as a histogram or frequency distribution and study its shape. A frequency distribution shows how often you recorded similar values for time.

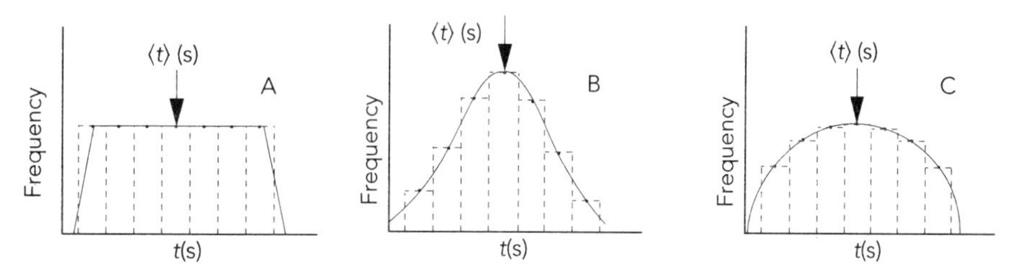

Fig. 2.7. Possible histogram shapes.

Suppose you dropped a ball from a tall building 54 times and recorded the following: no falls between 20.0 and 20.9 seconds, 6 different falls with times between 21.0 and 21.9 seconds, 6 different falls between 22.0 and 22.9 seconds and so on as shown in the table below. Then the frequency distribution or histogram would look like the one shown in diagram A in Figure 2.7.

Fall Times	Frequency
20.0–20.9	0
21.0–21.9	6
22.0–22.9	6
23.0–23.9	6
24.0–24.9	6
25.0–25.9	6
26.0–26.9	6
27.0–27.9	6
28.0–28.9	6
29.0–29.9	6
30.0–30.9	0

2.7.1. Activity: Frequency Distribution Prediction

Which of the diagrams in Figure 2.7 (A, B, or C) might have the shape you expect to find by drawing a frequency distribution of your data? Explain the reasons for your prediction.

HOW TO PLOT A FREQUENCY DISTRIBUTION

Since a frequency distribution of the fall-times shows how many times you recorded each time, you can draw this distribution by organizing your data as follows:

1. Load your spreadsheet file with the data to be plotted into the computer memory. (See Appendix A for details.)
2. Sort the column of data from the lowest time to the highest time. (See Appendix A for details.)
3. Count the frequency of occurrence of each quantity that was recorded. For example, if you recorded a time of .45 seconds five different times, the frequency of .45 seconds is 5.
4. The horizontal axis of your graph indicates the quantities whose frequencies you are graphing; the vertical axis of your graph gives the frequencies. Above each quantity on the horizontal axis, draw a rectangle whose height corresponds to the frequency of that quantity. Repeat this step for each quantity measured.

As an example, consider a very simple frequency distribution. Imagine that you have caught 10 fish. Of these 10, four are 3" long, two are 4" long, and four are 5" long. The frequency distribution would appear as follows:

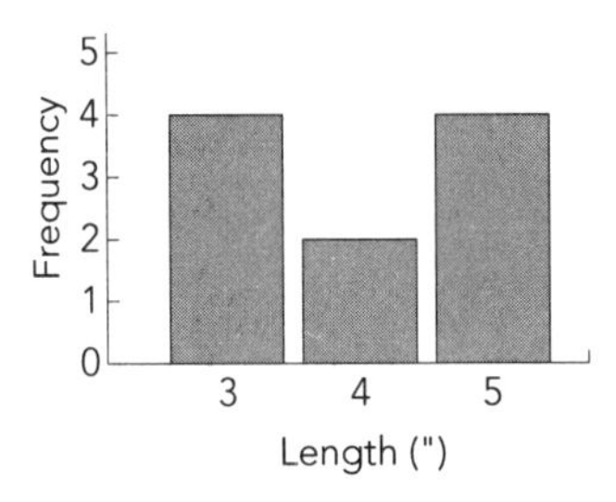

Fig. 2.8. A sample frequency distribution.

2.7.2. Activity: Frequency Distribution of Your Data

a. Draw a frequency distribution diagram (known as a histogram) of your time-of-fall data for the ball in the grid below.

t(s)	Freq.	t(s)	Freq.

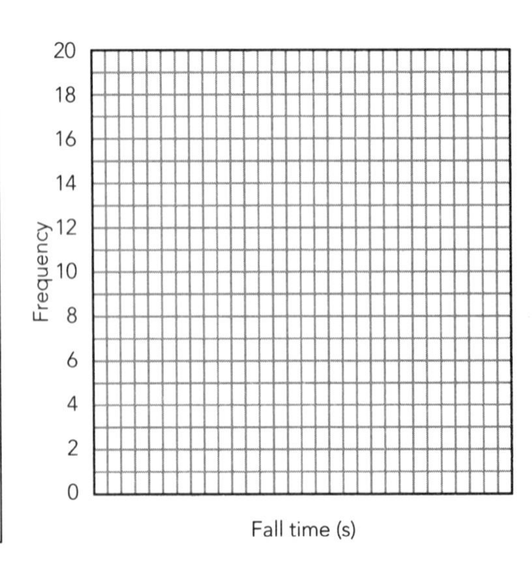

b. Next, using a different color pen or pencil, sketch in the results of the rest of the class in the histogram above. How does the shape of the class frequency distribution above compare with Figure C.2 in Appendix C? Does the variation in the time-of-fall data seem "normally distributed"? How does it compare to your prediction in Activity 2.7.1? Explain.

As you will observe later, a normal distribution of variation in a series of measurements can lead to a bell-shaped curve when the variations in measurement are the result of a number of smaller variations that occur randomly from measurement to measurement. Although the underlying events are random, and hence unpredictable, the nature of the variation becomes predictable. Puzzled? Stay tuned, we'll tackle this idea again.

2.8. SYSTEMATIC ERROR–HOW ABOUT THE ACCURACY OF YOUR TIMING DEVICE AND TIMING METHODS?

As the result of problems with your measuring instrument or the procedures you are using, each of your measurements may tend to be consistently too high or too low. If this is the case, you probably have a source of systematic error. There are several types of systematic error.

Most of us have set a watch or clock only to see it gain or lose a certain amount of time each day or week. In ordinary language we would say that such a time-keeping device is inaccurate. In scientific terms, we would say that it is subject to systematic error. In the case of a stopwatch or digital timer that doesn't run continuously like a clock, we have to ask an additional set of questions. Does it start up immediately? Does it stop exactly when the event is over? Is there some delay in the start and stop time? A delay in starting or stopping a timer could also cause systematic error.

Finally, systematic error can be present as a result of the methods you and your partner are using for making the measurement. For example, are you starting the timer exactly at the beginning of the event being measured and stopping it exactly at the end? Are you dropping the ball from a little above the exact starting point each time? A little below?

Fig. 2.9.

It is possible to correct for systematic error if you can quantify it. Suppose that God, who is a theoretical physicist, said that the distance in meters, y, that a ball falls after a time of t seconds near the earth's surface in most places is given by the equation

$$y = \frac{1}{2} a_g t^2$$

where a_g is the gravitational constant (equal to 9.8 m/s/s). (In this idealized equation the effects of air resistance have been neglected.)

Does the theoretical value for the time-of-fall lie within the standard error of your average measured value? In the activity that follows, you should compare your average time-of-fall with that expected by theory. If you determine that a systematic error probably exists, can you devise a way to determine its cause and magnitude?

2.8.1. Activity: Is There Systematic Error in the Data?

a. Using a 20-m distance of fall, calculate the theoretical, God-given, time-of-fall in the space below.

b. Does the theoretical value lie in the range of *your own* average value with its associated uncertainty? If not, you probably have a source of systematic error.

c. If you seem to have systematic error, explain whether the measured times tend to be too short or too long and list some of the possible causes in the space below.

d. Devise a method to find the causes of your systematic error. Explain what you did in the space below.

NORMAL DISTRIBUTIONS AND NUCLEAR DECAY

2.9. THE RANDOM OR DRUNKARD'S WALK IN ONE DIMENSION

Let's return to the question of how the accumulation of many small random events can lead to a pattern of variations in a series of measurements that is "normally distributed" and thus has a histogram that looks like a bell-shaped curve. To do this we are going to consider a series of measurements on the locations of drunkards in an alley. If the drunkards leave a bar in the center of the alley and then make many small steps at random, where will we find them sleeping in the morning? Assume that during each second of time that passes a drunk has an equal probability of staggering to the right, standing still, or staggering to the left.

Fig. 2.10.

2.9.1. Activity: The Random Walk–Predictions

a. If a drunk leaves the bar and staggers around taking one-foot long steps to the right and to the left at random, what will the drunk's average distance in feet from the bar be after he has had a chance to take many steps?

b. Is it more likely that a drunk will be close to the bar or far away? Why?

c. Suppose a drunk has had enough time to take 20 steps. Is it *possible* to find the drunk 20 steps from the bar in either direction? Is it probable? Explain!

Simulating the Drunkard's Walk with Dice

Let's simulate the drunkard's walk. In this exercise you are to come out of the bar drunk and try to take 20 steps at random. Where are you after attempting 20 steps? Each possible step is analogous to a source of variation in a measurement in the physics laboratory. For example, in dropping a ball there might be 20 variables that would affect its rate of fall: for example, air currents, inconsistent timing, the hand quivering upon release, etc., etc. For this simulation your group will need:

- 1 six-sided die

Recommended group size:	2	Interactive demo OK?:	N

2.9.2. Activity: The Random Walk–Simulated Observations

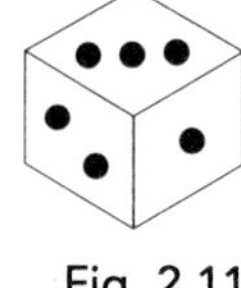

Fig. 2.11.

a. Imagine that you have just come out of a bar at the center of a narrow alley. Roll your die. For 1 or 2 stagger one step to the left. For 3 or 4 just twirl around in the same place. For 5 or 6 take a step to the right. Now where are you? Starting from the new location roll the die again to take another step. Do this a total of 20 times. Where are you after 20 tries? Eight to the left? Three to the right? etc. Repeat the procedure once more and enter your data below.

Step No.	No. on die	Step (−1, 0, +1)	New pos.
1.			
2.			
3.			
4.			
5.			
6.			
7.			
8.			
9.			
10.			
11.			
12.			
13.			
14.			
15.			
16.			
17.			
18.			
19.			
20.			

Step No.	No. on die	Step (−1, 0, +1)	New pos.
1.			
2.			
3.			
4.			
5.			
6.			
7.			
8.			
9.			
10.			
11.			
12.			
13.			
14.			
15.			
16.			
17.			
18.			
19.			
20.			

b. For each "walk" you just took, mark your final position on the histogram. Also mark the final position of other drunks in your class.

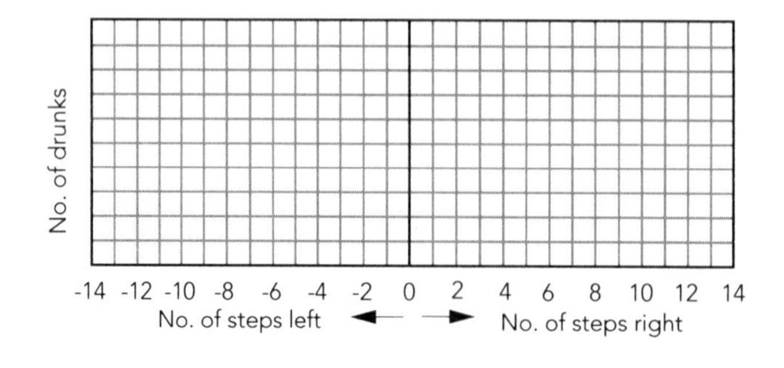

c. Does the variation in the data look as if it will be "normally distributed" when each value has an uncertainty that results from the accumulation of many random unpredictable steps?

Simulating the Drunkard's Walk with a Spreadsheet

Using a spreadsheet to help you take your imaginary walks from the bar is much faster than rolling a die. Most spreadsheets have a built-in random function that generates a random number greater than or equal to zero and less than 1. By manipulating this function and finding the integer value of the resulting numbers, it is easy to generate an integer with values of –1, 0, and +1. The equations that work for a typical spreadsheet are shown below.

	A	B	C
1	Step	Random 0 to 1	Random -1,0,+1
2	1	=RAND()	=INT(3*B2-1)
3	2	=RAND()	=INT(3*B3-1)
4	3	=RAND()	=INT(3*B4-1)
5	4	=RAND()	=INT(3*B5-1)
6	5	=RAND()	=INT(3*B6-1)
7	6	=RAND()	=INT(3*B7-1)
8	7	=RAND()	=INT(3*B8-1)
9	8	=RAND()	=INT(3*B9-1)
10	9	=RAND()	=INT(3*B10-1)
11	10	=RAND()	=INT(3*B11-1)
12	11	=RAND()	=INT(3*B12-1)
13	12	=RAND()	=INT(3*B13-1)
14	13	=RAND()	=INT(3*B14-1)
15	14	=RAND()	=INT(3*B15-1)
16	15	=RAND()	=INT(3*B16-1)
17	16	=RAND()	=INT(3*B17-1)
18	17	=RAND()	=INT(3*B18-1)
19	18	=RAND()	=INT(3*B19-1)
20	19	=RAND()	=INT(3*B20-1)
21	20	=RAND()	=INT(3*B21-1)
22			
23		Sum of Steps	=SUM(C2:C22)

By entering these equations into a spreadsheet and recalculating, you can simulate multiple 20-second walks. Each simulated walk should take a mere fraction of a second. To recalculate, press the command key and the equal sign (Mac) or the F9 key (Windows).

2.10. NATURAL RADIOACTIVITY AND STATISTICS

Fig. 2.12.

What happens when the particles coming from radioactive materials are counted during a time interval such as a second? What variation might we expect in repeated measurements? In particular, *does the shape of the frequency distribution of the number of counts per second look like that of the repeated measurements of time for a falling ball?*

Before exploring these questions, let's briefly review radioactivity. Radioactivity is understood as a phenomenon in which every once in a while a nucleus in a collection of radioactive atoms "decays" by ejecting either a gamma ray, a beta particle, or an alpha particle. Radioactivity is a statistical process in which a series of slight disturbances of a nucleus lead to a decay.

Heavy elements such as uranium and thorium occur naturally in rocks and soil. Even a few tiny grains of such an element can contain hundreds of billions of nuclei that are radioactive. The tiny particles ejected from a sample of radioactive matter can be counted by an electronic device known as a Geiger tube.

Nuclear Counting

If a given radioactive nucleus lives on the average for billions of years before undergoing decay, then a collection of many such nuclei will appear, on the average, to give off the same number of particles each second. What will happen if you count the radiation coming into a Geiger tube for one second and then repeat the measurement 20 times, 100 times . . . 1000 times? Do you expect to see variations?

In the next activity, you will count the number of beta particles coming into a Geiger tube during a fixed time interval. Then you'll do it again and again and again and again, etc. You can use your measurements to find the average value of your counts per time interval, calculate the standard deviation, and produce a graph of the frequency distribution.

The main point of the experiment and its analysis is to compare the shape of the frequency distribution curve for beta particles per time interval to that for falling balls. Are they similar? What happens when you take thousands of "data points?" What does the shape of the frequency distribution look like then?

For this experiment your group will need a radioactive source, a Geiger tube, and a computer-based laboratory system that can keep track of particles coming into a Geiger tube automatically. This allows you to take large quantities of data painlessly and thus continue your exploration of the characteristics of repeated measurements on quantities subject to random fluctuations.

Measuring Counts per Time Interval

In the exploration of the statistics of radioactivity, you can use a computer-based laboratory set up as a radiation counting system. Your group will need the following equipment:

- 1 computer-based laboratory system
- 1 radiation sensor
- 1 counting software
- 1 radioactive source (low level)

Recommended group size:	2	Interactive demo OK?:	Y

Note: If you don't have computer-based radiation monitoring equipment, the repeated counts can be taken with an old-fashioned Geiger tube and scalar system.

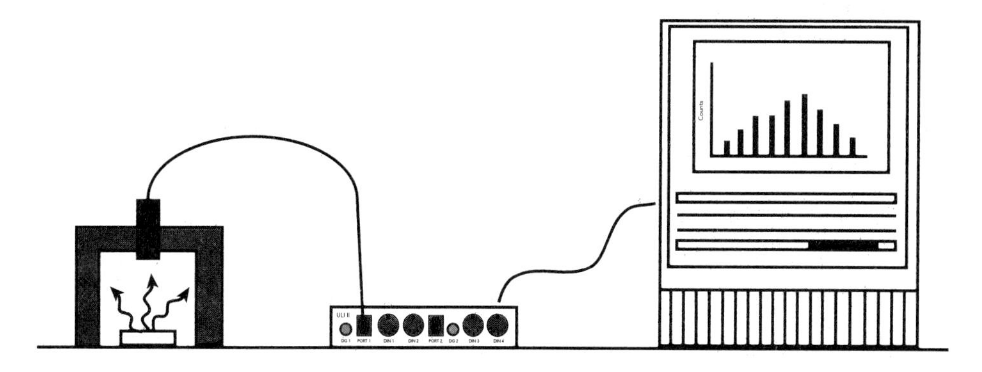

Fig. 2.13. A computer-based laboratory radiation monitoring system.

Getting Used to the Radiation Monitoring Apparatus

1. Refer to the radiation software manual for details on how to use the system with your microcomputer.
2. Set up the system and play around with the features of the event counting software.
3. See what happens when you move the source farther away from the Geiger tube in the radiation sensor.
4. Figure out how *to change the counting interval* and *to display counts/interval vs. time in a graph* and *in a table* on the computer screen.

Doing the Experiment

First you will be determining the counts/interval for 20 intervals and analyzing the data statistically in the same way you analyzed the time-of-fall data for the ball. Next you will let the computer plot the frequency distribution histogram for you automatically as it collects the data. This will allow you to obtain a frequency distribution for several thousand repeated counts/interval measurements. This in turn will enable you to tell whether or not the bell-shaped curve is a reasonable shape for the frequency distribution.

Set up the monitoring system and adjust the distance between the radioactive source and the Geiger tube or the time interval for the counts until the average number of counts/interval is about 20. You can select this time interval to be 1/20th of a second, 1/10th of a second, 1/2 of a second, 1 second, etc. Once the distance from source to detector is adjusted to give about 20 counts in the chosen time interval, *the source and the radiation sensor should not be bumped or disturbed.*

2.10.1. Activity: Is There Random Variation in the Nuclear Counting Data?

a. Count the radiation coming through the Geiger tube for the preset time interval you have chosen. Now, repeat the measurement 20 times or more using the same time interval. Record how many times you got 0 counts per interval, 1 count per interval, and so on. Then plot the frequency distribution of your results below.

Trial	Counts/ Interval	Trial	Counts/ Interval
1		11	
2		12	
3		13	
4		14	
5		15	
6		16	
7		17	
8		18	
9		19	
10		20	

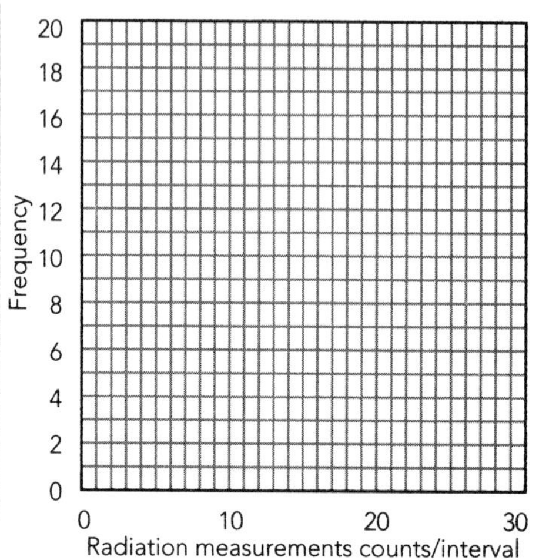

b. Determine the average counts/interval and standard deviation from the computer screen or, if necessary, use a spreadsheet or scientific calculator to determine the standard deviation. (There are many ways to skin a cat!) List the values in the space below.

c. Determine the percentage of your data points that lie within $\pm\sigma_{sd}$ of the average. Show your calculations below.

d. How does this percentage compare with what you found for the falling ball data? What is the "practical meaning" of the standard deviation, σ_{sd}, for the nuclear radiation data?

e. Finally, take advantage of the power of the computer-based radiation monitoring system and set the computer to monitor several thousand repeated counting intervals with an average count of about 20. (Please appreciate the fact that this would take you hours and hours to do manually!) The frequency distribution for this large number of data points can be displayed on the computer screen automatically. Follow the instructions *Saving and Printing Graphs and Tables* in Appendix B to print out a copy of your huge frequency distribution and affix it in the space below.

f. Study and comment on the shape of the resulting histogram. Does it look like a bell-shaped curve? Does nuclear counting seem to have a random variation? Explain.

2.11. CONFIDENCE INTERVALS AND REPORTING UNCERTAINTY

Standard Deviation of the Mean

To get a good estimate of some quantity you need several measurements, and you really want to know how uncertain the *average* of those several measurements is, since it is the average that you will write down (as a best estimate). This uncertainty in the average is known as the *standard deviation of the mean* or SDM for short.

It is this quantity that answers the question, "If I repeat *the entire series of N measurements* and get a second average, when do I have a 68% confidence that this second average will come close to the first one?" The answer is that you should expect a second average (that results from redoing the set of measurements) to have a 68% probability of lying within one SDM of the first average you determined. Thus, the SDM is sometimes referred to as a 68% confidence interval.

Once you know the Standard Deviation (σ_{sd}), it is simple to calculate an estimate for the standard deviation of the mean or SDM. This is simply the standard deviation of the sample of N measurements divided by the square root of N.

$$\text{SDM} = \frac{\sigma_{sd}}{\sqrt{N}}$$

It is also referred to at times as the standard error. Since the SDM is actually a measure of *uncertainty* rather than of an error (in the sense of a mistake), we prefer not to use the term standard error. The next activity can be done with either a spreadsheet to a scientific calculator. Note that all of the questions in this activity refer to the data you collected in Activity 2.10.1.

2.11.1. Activity: Calculating the SDM

a. How many times were counts recorded with the help of the computer?

$N =$

b. State the average counts/interval below.

c. The computer reported the Standard Deviation, σ_{sd}, to be

d. Use the standard deviation reported by the computer to calculate the Standard Deviation of the Mean (i.e., the SDM).

e. I am confident on the 68% level that if I repeated Activity 2.10.1 that I would obtain an average that is within

_______ ± __________

of the average we obtained.

f. Explain in your own words how the standard deviation (σ_{sd}) differs from what the standard deviation from the mean (SDM) tells you.

Name ________________ Section ________ Date ________

UNIT 3: ONE-DIMENSIONAL MOTION I

A GRAPHICAL DESCRIPTION

This False Vampire bat navigates with spectacular speed and agility by emitting a series of supersonic calls that echo back and warn her of obstacles. When interfaced to a laboratory computer, how can this type of navigation system help you track your motion along a line and begin to learn about how motion is described by physicists?

UNIT 3: ONE-DIMENSIONAL MOTION I

A GRAPHICAL DESCRIPTION

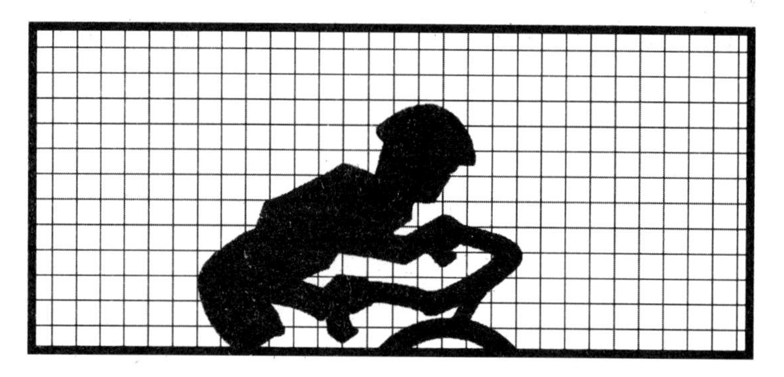

A picture is worth a thousand words!

OBJECTIVES

1. To learn about three ways that physicists can describe motion along a straight line–words, pictures, and graphs.

2. To acquire an intuitive understanding of position, speed, velocity, and acceleration.

3. To recognize how graphs can be used to describe changes in position, velocity, and acceleration of an object moving along a straight line.

4. To be able to use mathematical definitions of average velocity and acceleration in one dimension to determine these quantities from fundamental measurements.

3.1. OVERVIEW

A moving object might change its direction or speed as time passes. In order to use physics to describe such motions, you must learn how to describe them both graphically and mathematically. The study of the representation of motions using mathematical equations and graphs is known as *kinematics.*

Describing the motion is not always easy. For example, a cloud in the sky could be changing its size and shape as it moves. Physicists often start investigations by using simplifying assumptions. At first, you will concentrate on describing the motions of objects that are small or don't change shape. Initially you will pretend that objects are mathematical points, even when they are not! And for now, we will only study *one-dimensional* motions confined to a straight line.

To undertake the activities in this unit you will use a *computer-based laboratory system* consisting of a microcomputer, an electronic interface, a sensor, and special software to help you study motion. When you attach an ultrasonic motion sensor to this system you can automatically record the distance of your own body and other objects from the sensor. Computer software allows you to display how these distances change over time almost instantly in the form of a graph. In addition the software can use a programmed set of rules to calculate two other quantities that are used to describe motion from distance and time measurements. These include *velocity* and *acceleration.*

Since the actual rules for determining and graphing velocity and acceleration from distance and time measurements are programmed into the motion software, *you have a unique opportunity to develop an intuitive understanding of the concepts of velocity and acceleration before considering their formal mathematical definitions.*

This unit ends with an exploration of the formal mathematical definitions of average velocity and acceleration to represent "rate of motion" and the change in the "rate of motion" respectively.

DESCRIBING MOTION WITH WORDS AND GRAPHS

3.2. DESCRIBING POSITION CHANGES WITH WORDS AND GRAPHS

The activities in this first section on kinematics will help you learn to position changes using *words* and *graphs*. Activities in the next section will involve descriptions of changes in the velocity of an object.

Fig. 3.1. Describing motion with words: the person on the street versus the physics student.

Fig. 3.2. Describing motion with graphs: the person on the street versus the physics student.

For the activities in this section your group will need:

- 1 ruler
- 1 computer-based laboratory system
- 1 ultrasonic motion sensor
- 1 motion software
- 1 roll of masking tape (optional, for a number line)
- 1 meter stick (optional, for a number line)

Recommended group size:	3	Interactive demo OK?:	N

The Definition of Position Along a Line

Position and time are the two most fundamental measurements in the study of motion. Physicists define the *position* of an object that lies on a line as the distance between a point of interest on the object and some reference point or origin also lying on that line. Usually position of an object along a horizontal line is negative if it lies to the left of a reference point and positive if it lies to the right of the reference point. The convention for reference points and objects lying along a vertical line, position is positive if the object is above the reference point and negative if the object is below the reference point.

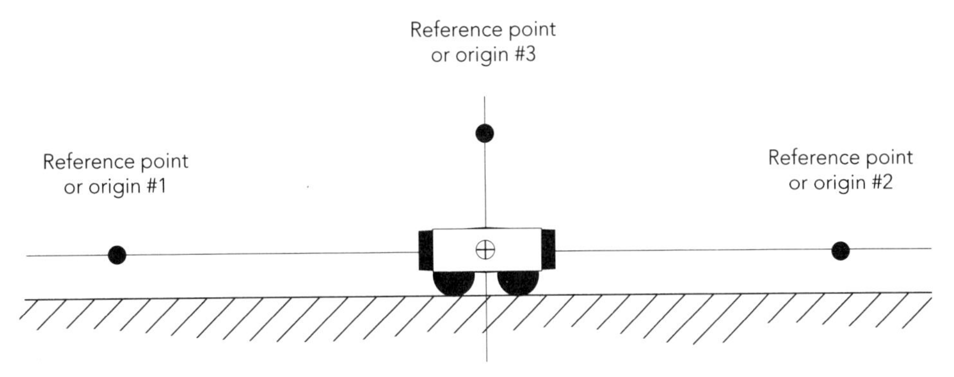

Fig. 3.3. A diagram of an object and various origins that can be used for finding position.

In this next activity you can use a ruler to measure distances in centimeters and apply the definition of position to find the positions of the cart relative to various reference points.

3.2.1. Activity: Finding Positions

a. Find the position of the center of the cart relative to origin #1. Do the cart and the reference point lie along a horizontal or vertical line? Is the position positive (+) or negative (−)?

b. What is the position of the cart's center relative to origin #2? Do the cart and reference point lie along a horizontal or vertical line? Is the position positive (+) or negative (−)?

c. What is the position in centimeters of the cart's center relative to origin #3? Do the cart and reference point lie along a horizontal or vertical line? Is the position positive (+) or negative (−)?

The Ultrasonic Motion Sensor

In the rest of the activities in this unit you will be using a computer-based laboratory system with an ultrasonic motion sensor and motion software. It is helpful to understand some of the basic characteristics of motion sensors to use them intelligently.

The ultrasonic motion sensor acts like a stupid bat when hooked up with a computer-based laboratory system. It sends out a series of sound pulses that are too high frequency to hear. These pulses reflect from objects in the vicinity of the motion sensor and some of the sound energy returns to the sensor. The computer is able to record the time it takes for reflected sound waves to return to the sensor and then, by knowing the speed of sound in air, figure out how far away the reflecting object is. There are several things to watch out for when using a motion sensor.

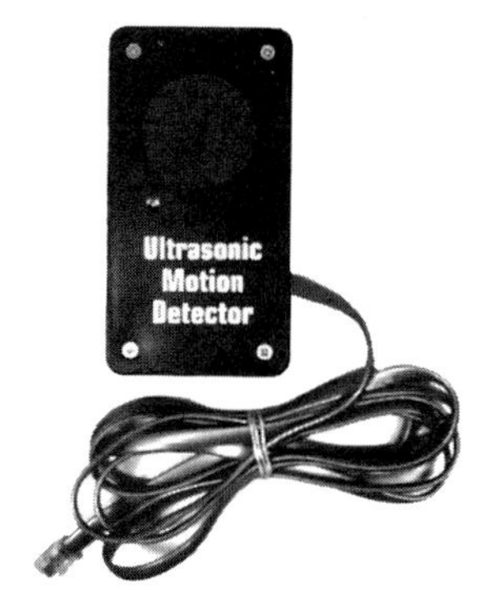

Fig 3.4. One of several models of the type of motion sensor used in this Activity Guide.

When Using a Motion Sensor

1. Do not get closer than 0.5 meters from the sensor because it cannot record reflected pulses that come back too soon after they are sent.
2. The ultrasonic waves spread out in a cone of about 15° as they travel. They will "see" the closest object. Be sure there is a clear path between the object whose motion you want to track and the motion sensor.
3. The motion sensor is very sensitive and will detect slight motions. You can try to glide smoothly along the floor, but don't be surprised to see small bump representing your steps in velocity graphs and even larger bumps in acceleration graphs.
4. Some objects like bulky sweaters are good sound absorbers and may not be "seen" very well by a motion sensor. You may want to hold a book in front of you if you have loose clothing on.

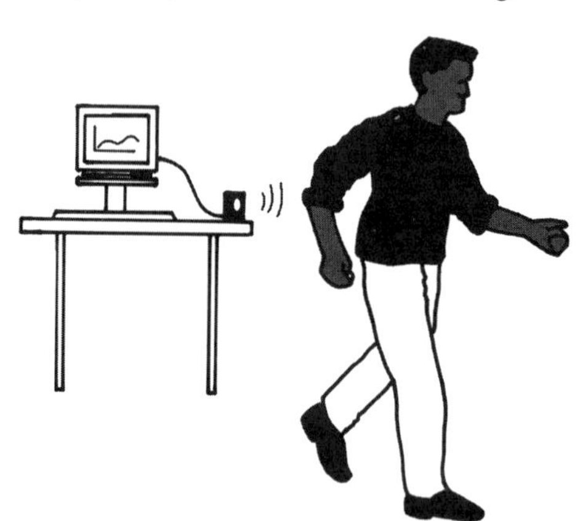

Fig. 3.5. Walking in front of a motion sensor attached to a computer-based laboratory system.

Important note on saving your files: You may be asked to use the data collected with the motion software in some of the upcoming mathematical analysis activities. For each activity, please save the graph sets and associated data on your data disk. We recommend that you identify your files by the unit, section, and activity number for later reference along with your group initials. Thus, if Smith, Ricci, and Pawelski work together on Unit 3, section 3.3, activity 3.3.1, the file name might be, for example, A3_3_1 (SRP).

Position vs. Time Graphs of Your Motion

The purpose of the next activity in this session is to learn how to relate graphs of position as a function of time to the motions they represent.

How does a position vs. time graph look when you move slowly? Quickly? What happens when you move toward the motion sensor? Away? After completing the next few activities, you should be able to look at a position vs. time graph and describe the motion of an object. You should also be able to look at the motion of an object and sketch a graph representing that motion.

For these computer-based position measurements:

- The motion sensor is the origin—that is, the reference point for all position graphs.
- Since the computer cannot tell right from left, the position is always assumed to be positive when using a motion sensor.

To do the activity and those that follow you should make sure that: (1) an interface is plugged into a power source and connected to your computer, (2) the motion sensor is plugged into the interface, and (3) that motion software has been loaded into the memory of your computer. You should then set up the software to record position vs. time graphs for about 15 seconds. (The manual for your counting software will contain instructions for how to accomplish this).

3.2.2. Activity: Interpreting Position Graphs

a. Make various position-time graphs while standing still and for different *steady* walking speeds and directions. You may want to use the diagrams below to relate the graphs you observe with descriptions in words of your motions.

Sketch of graph | Description of motion

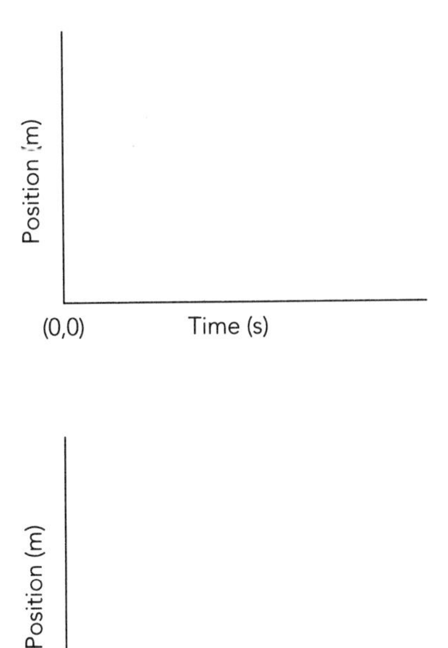

Sketch of graph	Description of motion

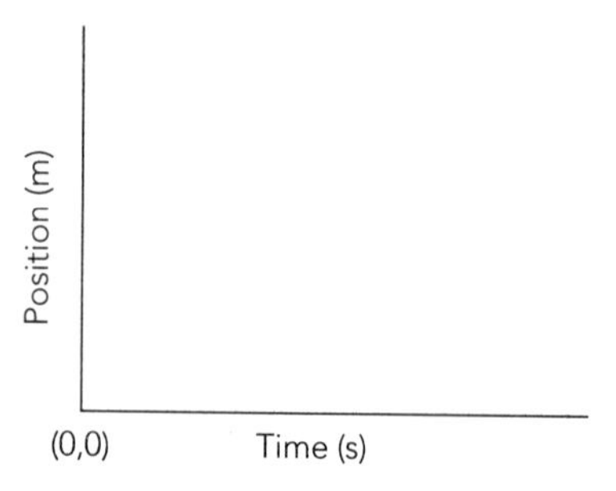

b. Describe the difference between the graph you made by walking away slowly and the one made by walking away more quickly.

c. Describe the difference between the graph made by walking toward and the one made walking away from the motion sensor.

Predicting the Shapes of Position vs. Time Graphs

A good way to verify that you understand how to interpret position vs. time graphs is to predict the shape of a graph of a set of motions that can be described in words. Then you can carry out the motions to verify your prediction.

3.2.3. Activity: Predicting a Position vs. Time Graph

a. Suppose you were to start 2.0 m in front of the sensor and walk away slowly and steadily for 6 seconds, stops for 3 seconds, and then walk toward the sensor quickly for 6 seconds. Sketch your prediction on the following axes using a *dashed* line.

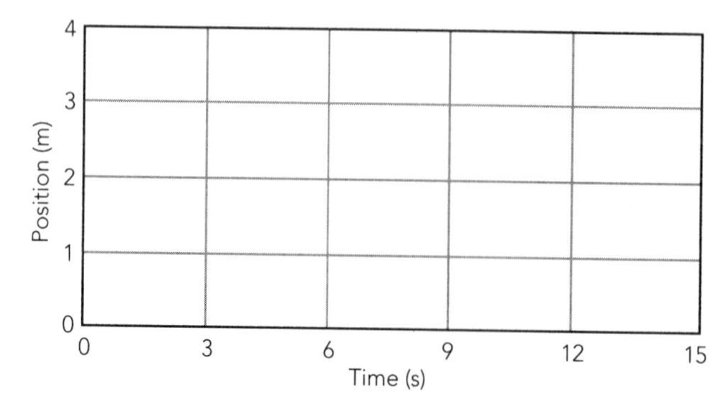

b. Test your prediction by attempting to produce a graph like that shown in the diagram above. Move in the way described and sketch the trace of your actual motion *with a solid line.*

c. Is your prediction the same as the final result? If not, describe how you would move to make a graph that looks like your *prediction.*

Matching Position vs. Time Graphs

Let's turn the activity you just did inside out. We'd like you to look at a graph of a completed motion and then describe the motion you think it depicts in words. Then you should be able to reproduce the motion you described. To do this you should close your eyes while your partner performs a set of motions consisting of a fairly simple combination of steady walking at *two different rates* of speed and standing still.

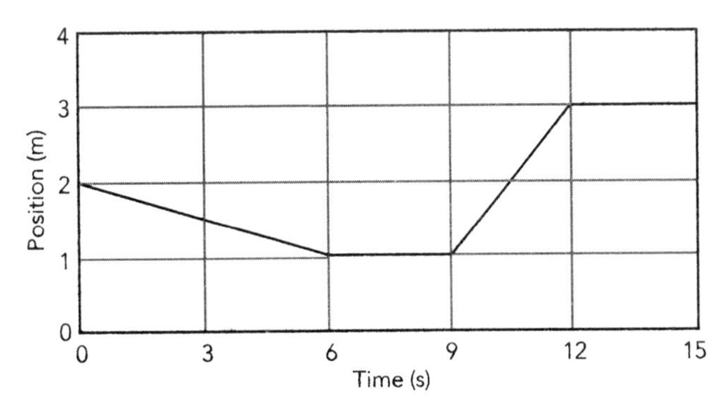

Fig. 3.6. Example of a graph depicting a combination of standing still with motion at two rates of speed.

3.2.4. Activity: Matching Position vs. Time Graphs

a. To do the first matching activity, ask your partner to begin with a blank graph and create a position graph while you look away. Sketch the shape of your partner's graph with a *dashed line* in the following graph and also save your partner's graph on the computer for comparison with your own. (See the manuals for the motion software for details on how to *collect two data sets for comparison.*)

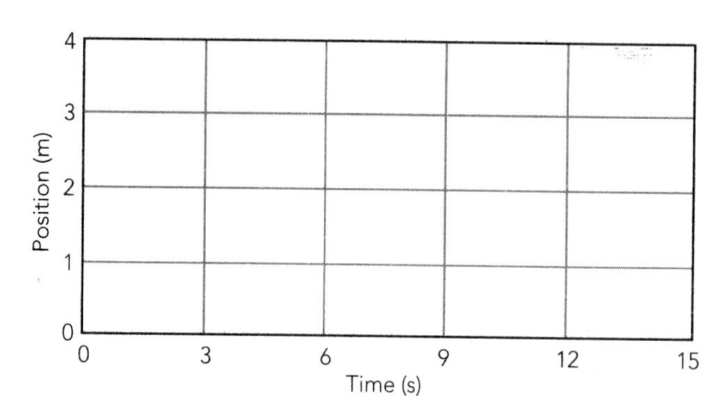

b. Describe in your own words how you plan to move in order to match your partner's graph.

c. Move to match your partner's graph on the computer screen. You may try this a number of times. It helps to work in a team. Get the times right. Get the positions right. Do this for yourself. (Each person in your group should do his or her own match.) *You will not learn very much by just watching!*

d. What was the difference in the way you moved to produce differently sloped parts of the graph you just matched?

e. Make curved position vs. time graphs like those shown below. **Note:** Before trying to reproduce these shapes, *start with a new blank graph.*

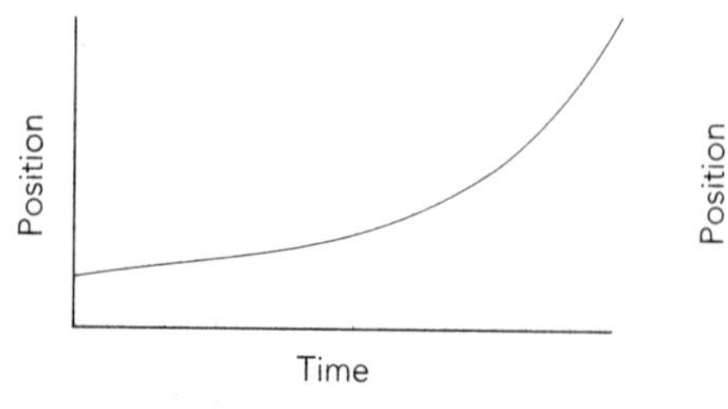

Fig. 3.7a. Graph 1.

Position

Time

Fig. 3.7b. Graph 2.

f. Describe how you must move to produce a position vs. time graph with each of the shapes shown.

Graph 1 answer:

Graph 2 answer:

g. What is the general difference between motions that result in a straight line position vs. time graph and those that result in a curved-line position vs. time graph?

3.3. DESCRIBING VELOCITY WITH WORDS AND GRAPHS

For the activities in this section your group will need:

- 1 ruler
- 1 computer-based laboratory system
- 1 ultrasonic motion sensor
- 1 motion software
- 1 motion software file

Recommended group size:	3	Interactive demo OK?:	N

Describing Speed and Direction Changes Graphically

You have already created position vs. time graphs for your body motions. Suppose you were a race car driver who is more interested in recording the *direction of the speed* of your car based originally on position measurements from a long-range motion sensor. How might you represent this on a graph?

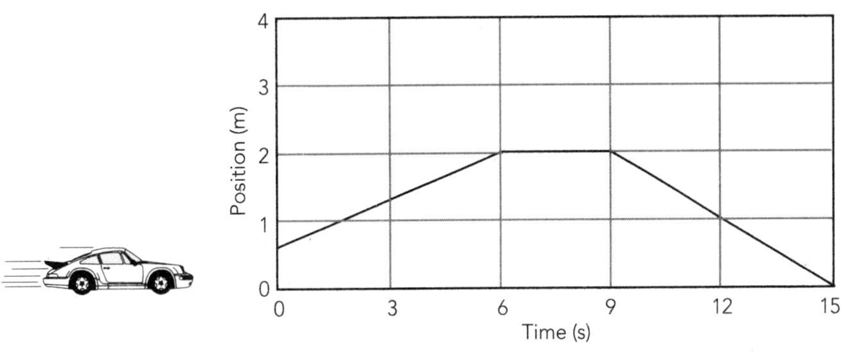

Fig. 3.8. An idealized graph depicting a race car standing still and moving at two different speeds in different directions.

After examining the graph in Figure 3.8, answer the following questions and see if you can devise a sensible method for representing speed and direction vs. time graphically.

3.3.1. Activity: Inventing a Speed/Direction Graph

a. What is the speed of the car between 0 and 2 seconds? What is its direction?

b. What is the speed of the car between 2 and 3 seconds? What is its direction?

c. What is the speed of the car between 3 and 5 seconds? What is its direction?

d. Try to devise a method for expressing both the speed and direction graphically. Use the graph frame below to help you sketch the graph.

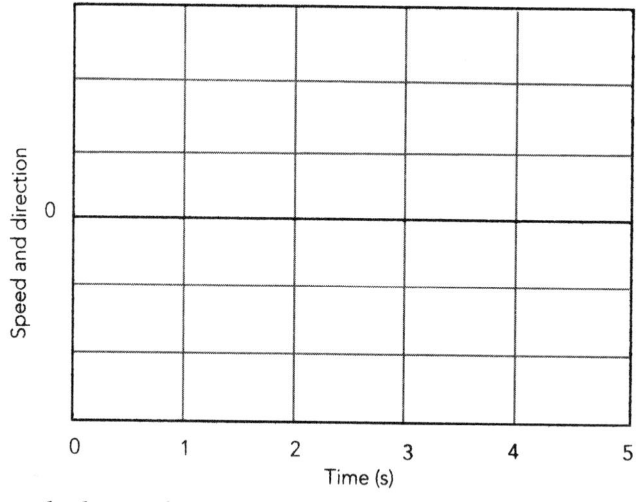

e. Explain the basis of your method. What rules did you come up with?

The Definition of Velocity and Velocity Graphs

Physicists have developed the concept of *velocity* to represent both the *direction* and *speed* of motion of an object. Basically, the velocity of an object moving along a line is a large number when that object has a high speed. The positive or negative sign that precedes the number signifies in what direction the object is moving. Motion-recording software can be set to display graphs of velocity vs. time in real time. The mathematical rules for calculating velocity are programmed into the motion software. In the next few activities you will observe the motion of your body and a cart and try to discover some characteristics of those rules.

Graphs of velocity vs. time are more challenging to create and interpret than those for position. A good way to learn to interpret them is to create and examine velocity vs. time graphs of your own body motions, as you will do in the next few activities.

For the next few activities you should open the motion software and set it to graph velocity. (To learn how to display other types of graphs, consult your manual for the motion software.) Set the *Velocity* axis from 21.0 to 11.0 m/s and the *Time* axis to read 0 to 5s.

3.3.2. Activity: Making Velocity vs. Time Graphs

Note: To *change the scale of your graph axis* so the trace fills the screen better, consult the manuals for motion software. Choose a different maximum and minimum velocity.

a. Make a velocity graph by walking *away* from the sensor *slowly and steadily.* Try again until you get a graph you're satisfied with and then sketch your result on the graph that follows. (We suggest you draw *smooth* patterns by ignoring smaller bumps that are mostly due to your steps.)

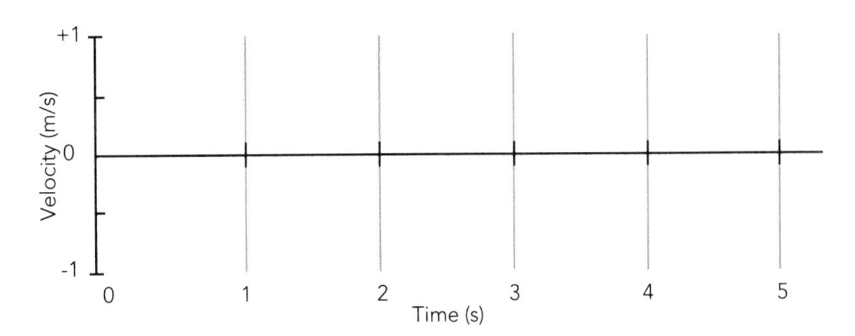

b. Make a velocity graph, walking *away* from the sensor *steadily* at a *medium speed.* Sketch your graph below.

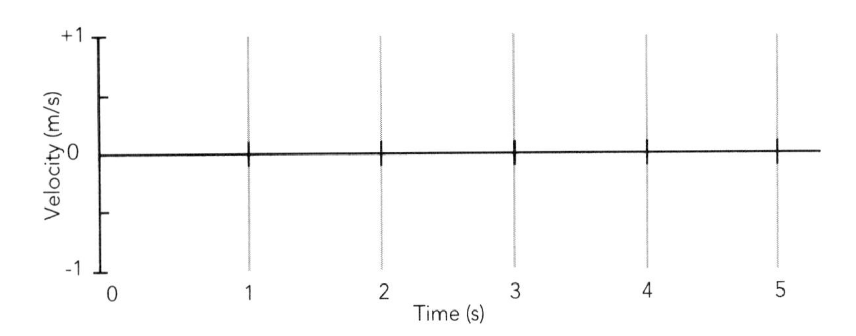

c. Make a velocity graph, walking *toward* the sensor slowly and *steadily.* Sketch your graph below.

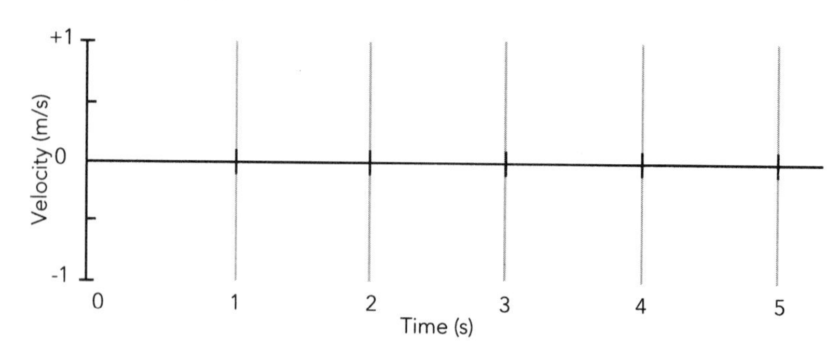

d. What is the most important difference between the graph made by *slowly* walking away from the sensor and the one made by walking away *more quickly?*

e. How are the velocity vs. time graphs different for motion *away* and motion *toward* the sensor?

Predicting Velocity vs. Time Graphs Based on Words

Suppose you were to undergo the following sequence of motions:

1. Walk away from the sensor slowly and steadily for 6 seconds.
2. Stand still for 6 seconds.
3. Walk toward the sensor steadily about twice as fast as before.

3.3.3. Activity: Predicting a Velocity vs. Time Graph

a. Use a *dashed line* in the following graph to record your *prediction* of the shape of the velocity graph that will result from the motion described above.

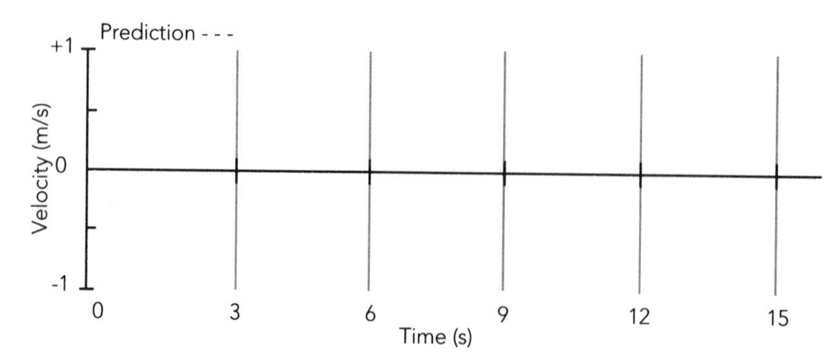

b. Compare predictions with your partner(s) and see if you can all agree. Use a solid line to sketch your group prediction in the graph above.

c. Adjust the time scale to 15 s in the Motion Software and then test your prediction. Repeat your motion until you are confident that it matches the description in words and then draw the actual graph on the axes below. Be sure the 6-second stop shows clearly.

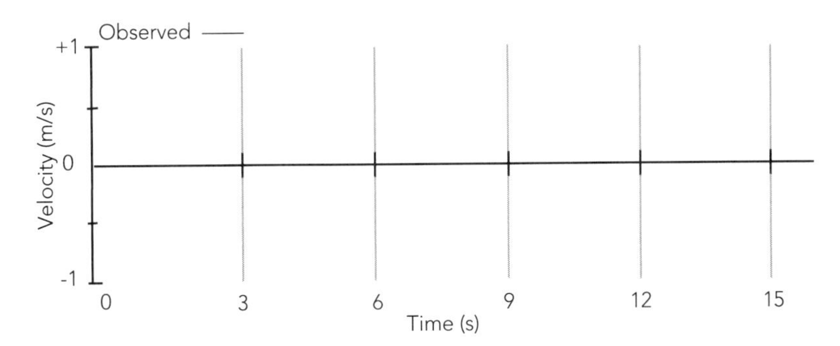

d. Did your prediction match your real motion? If not, what misunderstanding of what the elements of the graph represent did you have?

Velocity Vectors

As you have seen, for motion along a line (e.g., the positive x-axis) the sign (+ or −) of the velocity indicates the direction of motion. If you move away from the sensor (origin), your velocity is positive, and if you move toward the sensor, your velocity is negative. The faster you move away from the origin, the larger the positive number of the velocity. The faster you move *toward* the origin, the "larger" the negative number of the velocity.

Mathematicians define a quantity with magnitude and direction as a vector that can be represented by an arrow. Thus, velocity is a vector. The length of the velocity vector arrow is proportional to speed; the longer the arrow, the larger the speed. The direction of the arrow represents the direction of motion. Thus, if you are moving toward the right, your velocity vector can be represented by the arrow shown below.

On the other hand, if you are moving twice as fast toward the left, the arrow representing your velocity vector would be twice as long and point in the other direction as follows.

What is the relationship between a one-dimensional velocity vector and the *sign* of velocity? This depends on the way you choose to set the positive x-axis.

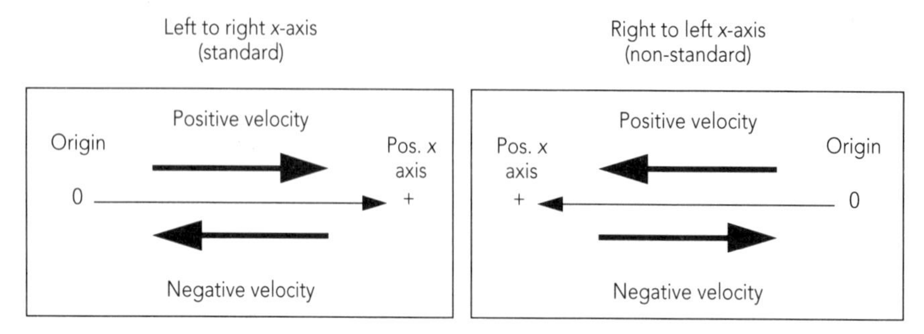

Fig. 3.9. Diagrams showing how positive and negative velocities are assigned for different directions of motion. The designation of positive or negative depends on how the x-axis is chosen. A direction of left to right for the x-axis is the most standard and should be used whenever possible.

In both the diagrams shown in Figure 3.9, the top arrows represent positive velocities. In the left diagram, the x-axis has been drawn so that the positive x-direction is toward the right. In the right diagram, the positive x-direction is toward the left. Thus, the top arrow still represents *positive* velocity. Likewise, in both diagrams the bottom arrows represent negative velocity even though they depict motion in opposite directions.

3.3.4. Activity: Sketching Velocity Vectors

a. Sketch the velocity vector representing walking quickly away from the sensor.

b. Sketch the velocity vector representing walking half as fast toward the sensor.

c. Sketch the velocity vector representing standing still.

Velocity Graph Matching

In the next activity, you will try to move to match a velocity graph shown on the computer screen. This is often much harder than matching a position graph as you did in the previous investigation. Most people find it quite a challenge at first to move so as to match a velocity graph. In fact, some velocity graphs that can be invented cannot be matched! To do this activity, pull down the File Menu and select *Open*. Then double click on the "Velocity Match" file. The following velocity graph should appear on the screen.

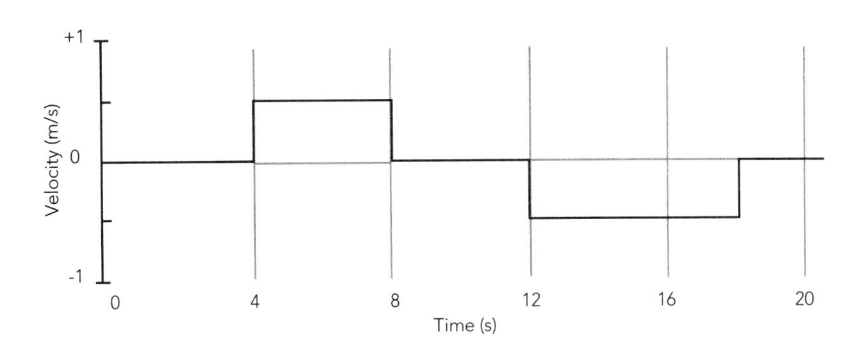

Fig. 3.10. Velocity graph.

3.3.5. Activity: Matching a Velocity Graph

a. Describe how you think you will have to move in order to match the given velocity graph.

b. Move in such a way that you can reproduce the graph shown. You may have to practice a number of times to get the movements right. Work as a team and plan your movements. Get the times right. Get the velocities right. You and each person in your group should take a turn. Then draw in your group's best match on the following axes.

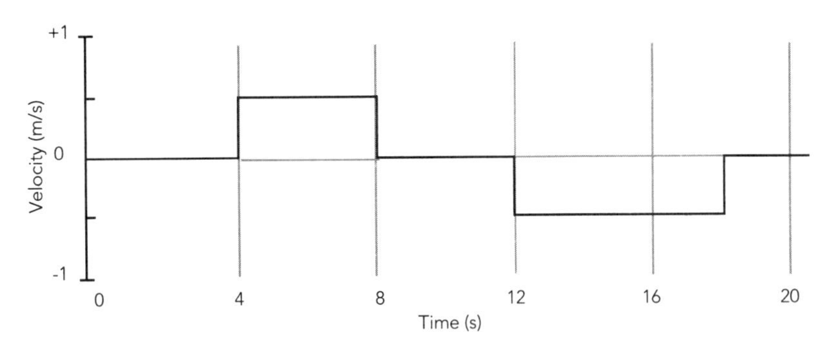

Fig. 3.11.

c. Describe how you moved to match each part of the graph.

d. Is it possible for an object to move so that it produces an *absolutely vertical* line on a velocity time graph? Explain.

e. Did you run into the motion sensor on your return trip? If so, why did this happen? How did you solve the problem? Does a velocity graph tell you where to start? Explain.

POSITION, VELOCITY, AND ACCELERATION GRAPHS

3.4. RELATING POSITION AND VELOCITY GRAPHS

Creating a Velocity Graph from a Position Graph

You have looked at position and velocity vs. time graphs separately as different ways to represent the same motion. It ought to be possible to figure out the velocity at which someone is moving by examining her/his position vs. time graph. Conversely, you ought to be able to figure out how far someone has traveled (change in position) from a velocity vs. time graph.

To explore how position vs. time and velocity vs. time graphs are related, you will need the following:

- 1 computer-based laboratory system
- 1 ultrasonic motion sensor
- 1 set of motion software
- 1 roll of masking tape (optional, for number line)
- 1 meter stick (optional, for number line)

Recommended group size:	3	Interactive demo OK?:	N

To complete the next activity, you'll need to set up the Motion Software to display both position vs. time and velocity vs. time simultaneously for a period of 5 seconds.

1. Open the motion software and set it up to display two graphs.
2. Set one graph to display Position from 0 to 4 m for 5 s.
3. Set the other graph to display Velocity from –1 to 1 m/sec for 5 s.

3.4.1. Activity: Velocity Graphs from Position Graphs

a. Carefully study the position graph shown below and predict the velocity vs. time graph that would result from the motion. Using a *dashed line*, sketch your *prediction* of the corresponding velocity vs. time graph on the velocity axes.

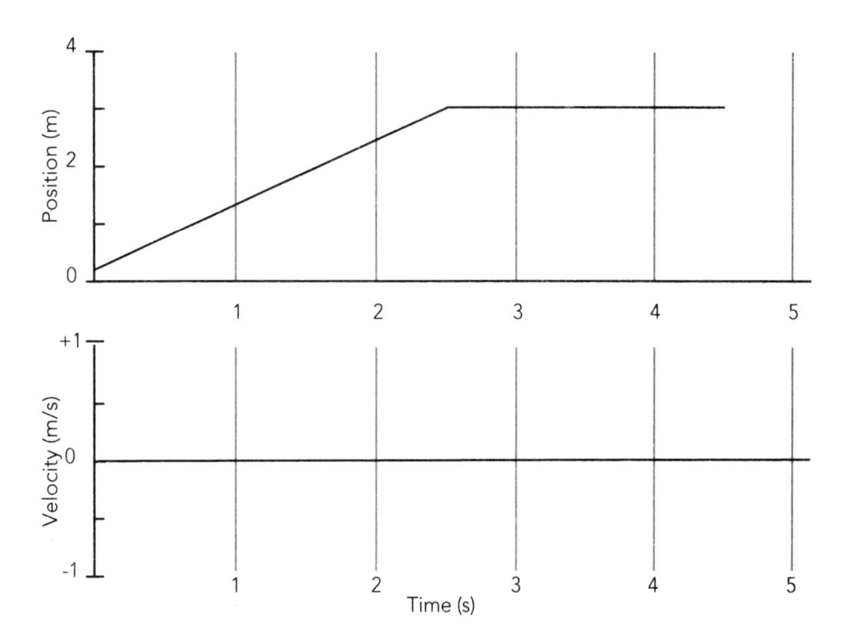

b. After each person in your group has sketched a prediction, test your prediction by using the motion software and sensor and walking to match the position vs. time graph shown. When you have made a good approximation of the position graph, sketch your actual graph on the position vs. time graph frame shown above.

c. Use a *solid line* to draw the actual velocity graph using the graph frame shown above (on the same graph as your prediction). (Do not erase your prediction.)

d. How would the position graph be different if you moved faster? Slower?

e. How would the velocity graph be different if you moved faster? Slower?

Important note: Please *save* the file for this activity since you may be asked to open it and make some calculations based on it in another session.

Creating a Position Graph from a Velocity Graph

The final challenge is to be able to produce position vs. time graphs from velocity graphs. To do this successfully, you must know the position of the person or object of interest at least one of the times.

3.4.2. Activity: Finding Position from Velocity

a. Carefully study the following velocity graph. Using a *dashed line,* sketch your *prediction* of the corresponding position graph on the bottom set of axes. (Assume that you started at 1 meter.)

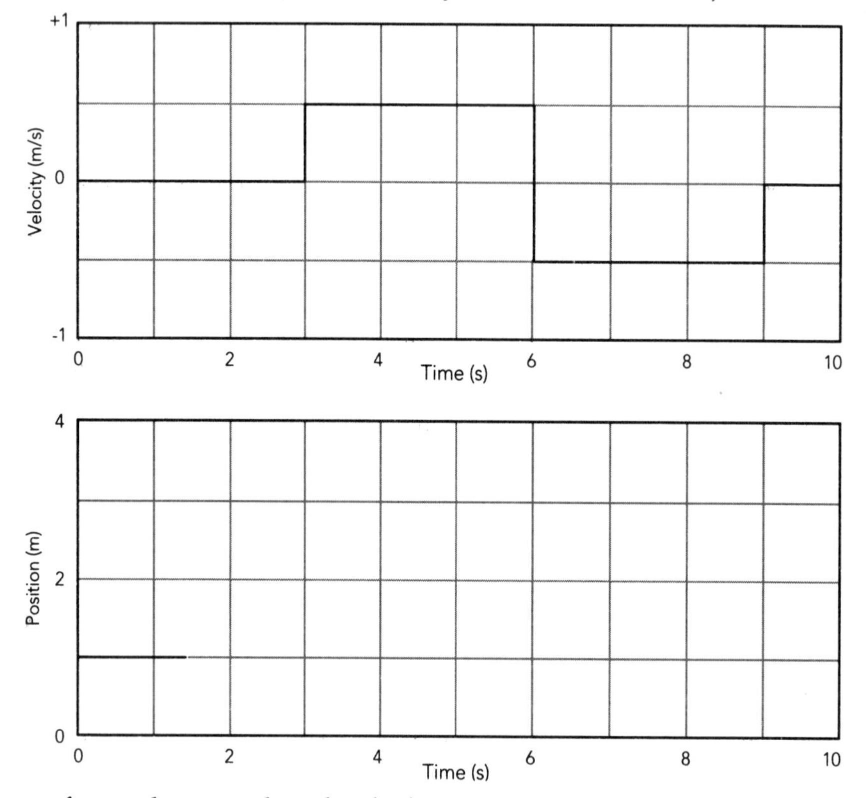

b. After each person has sketched a prediction, do your group's best to duplicate the top (velocity vs. time) graph by walking. (Reset the *Time* axis to 0 to 10 sec before you start.) When you have made a good duplicate of the velocity vs. time graph, draw your actual result over the existing velocity vs. time graph.

c. Use a *solid line* to draw the actual position vs. time graph on the same axes with your prediction. *Do not erase your prediction.*

d. How can you tell from a *velocity* vs. time graph that the moving object has changed direction?

e. What is the velocity at the moment the direction changes?

f. Is it possible to move your body (or an object) to make *vertical* lines on a position vs. time graph? Why or why not?

g. How can you tell from a position vs. time graph that your motion is steady (motion at a constant velocity)?

h. How can you tell from a velocity vs. time graph that your motion is steady (constant velocity)?

3.5. INTRODUCTION TO ACCELERATION

There is a third quantity besides position and velocity that is used to describe the motion of an object—acceleration. Acceleration is defined as *the rate of change of velocity with respect to time.* In this investigation you will begin to examine the acceleration of objects. Because of the jerky nature of the motion of your body, acceleration graphs are very complex. It will be easier to examine the motion of a cart. For the activities in this section you will need:

- 1 computer-based laboratory system
- 1 ultrasonic motion sensor
- 1 motion software
- 1 low-friction dynamics cart
- 1 ramp, 2 m

Recommended group size:	3	Interactive demo OK?:	Y

In the next two activities you will examine the velocity and acceleration of a low-friction cart just *after* it has been pushed. Later, you will examine the acceleration of a low-friction cart while it is being pushed. To graph the motion of a cart just after it is pushed, you can hold the cart with your hand and push it along a level ramp or a smooth level surface.

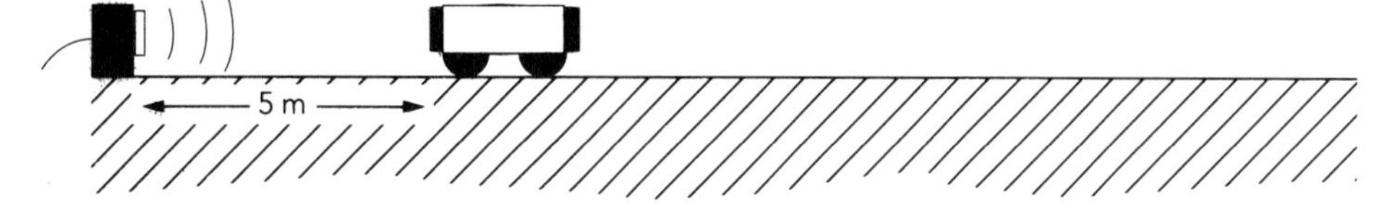

Fig. 3.12. Diagrams showing the ramp, cart, and motion sensor setup.

Position and Velocity Graphs of a Cart After a Push

Before examining an acceleration graph of our pushed cart, let's create position and velocity vs. time graphs.

1. Set up the motion sensor at the end of the ramp. If the cart has a friction pad, place it out of contact with the ramp so that the cart can move freely.
2. Open the motion software and set up position and velocity axes shown in the following graphs.
3. When taking measurement, give the cart a push away from the motion sensor and start collecting data with the motion software just as you finish pushing the cart.

3.5.1. Activity: Cart Motion After a Push

a. Based on your observations of your body motions, predict how the position and velocity graphs will look if the cart is pushed away from the motion sensor starting at the 0.5-meter mark? Sketch your predictions with dashed lines on the following axes.

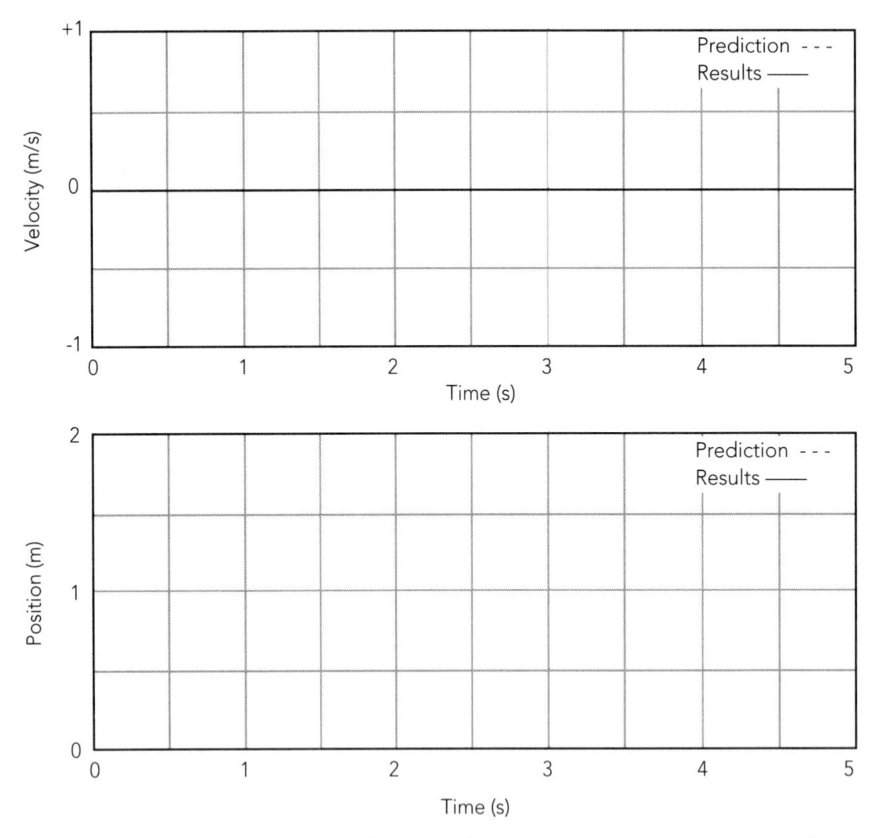

b. Test your prediction by collecting data on the moving cart with the motion software. Sketch your results with *solid lines* on the axes.

c. Did your position-time and velocity-time graphs agree with your predictions? What characterizes constant velocity motion on a position-time graph?

d. What characterizes constant velocity motion on a velocity vs. time graph?

An Acceleration Graph Representing Constant Velocity

You should have observed that the velocity of a low-friction cart just after it is pushed is approximately constant. Based on the definition of acceleration, what would an acceleration vs. time graph of the cart look like?

3.5.2. Activity: Graphing Acceleration vs. Time

a. What should an acceleration vs. time graph look like for the cart motion you just observed? Use the definition of acceleration to sketch a dashed line on the axes that follow.

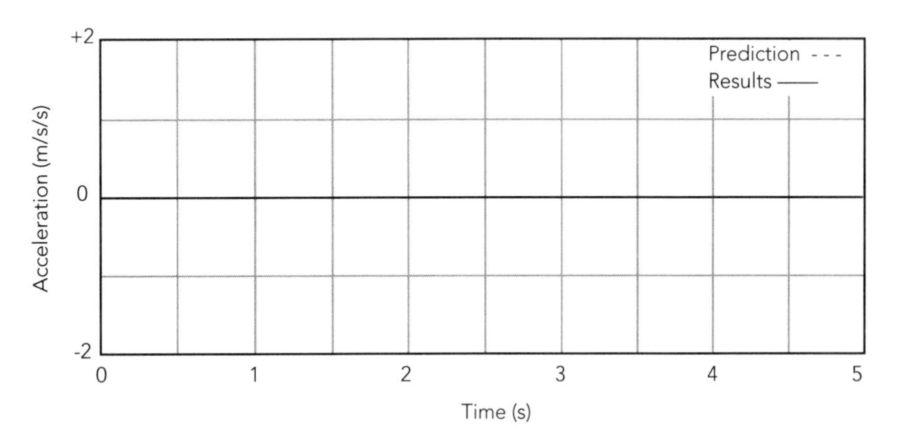

b. Change the graph axes to display the acceleration graph of the cart instead of the position graph, and then sketch the acceleration graph using a solid line on the axes above.

c. Does the acceleration vs. time graph you observed agree with your prediction?

Finding Accelerations Using Motion Diagram Vectors

To find the average acceleration of the cart during some time interval (the average time rate of change of its velocity), you must measure its velocity at two different times, calculate the difference between the final value and the initial value, and divide by the time interval.

An alternative way for finding an acceleration is to use a position-velocity motion diagram and then determine an acceleration vector. To do this you must first find the vector representing the *change in velocity* by subtracting the initial velocity vector from the final one. Then you divide this vector by the time interval.

For example, if a race car speeds up between time t_1 and t_2, from a velocity to a velocity , then the change in velocity is given by $\Delta v = v_2 - v_1$. And finally the acceleration is given by

$$\langle a \rangle = \frac{\Delta v}{\Delta t} = \frac{v_2 - v_1}{t_2 - t_1}$$

The use of vectors to depict how Δv can be found by vector subtraction is illustrated in the following diagram.

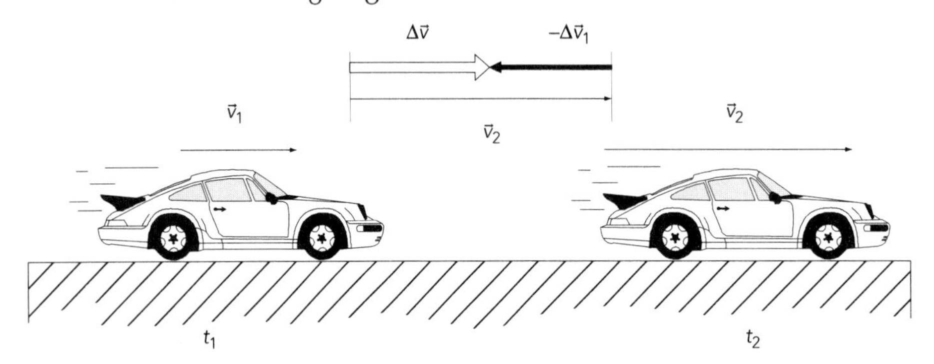

Fig. 3.13. Diagram showing how to use a motion diagram and a vector difference to find the change in velocity of the race car, Δv.

3.5.3. Activity: Using Vectors to Find Acceleration

a. The following diagram shows the positions of the cart at equal time intervals. (This is like taking snapshots of the cart at equal time intervals.) At each indicated time, sketch a vector above the cart that might represent the velocity of the cart at that time while it is moving at a constant velocity away from the motion sensor.

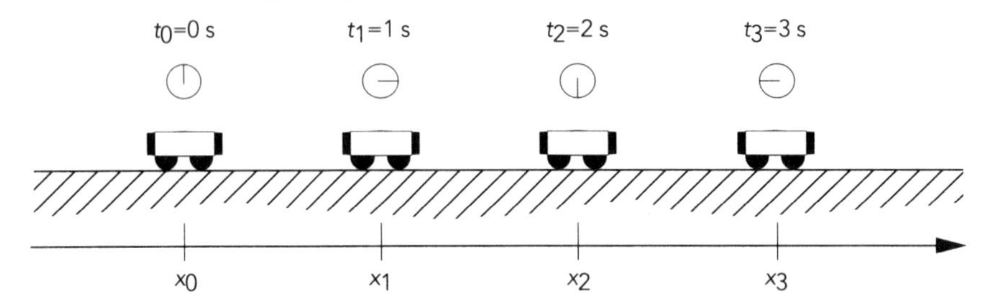

b. Explain how you would find the vector representing the change in velocity between the times 1.0 s and 2.0 s in the above diagram. From this vector, what value would you calculate for the acceleration? Explain. Is this value in agreement with the acceleration graph you obtained in Activity 3.5.2?

c. Does the acceleration vs. time graph you observed agree with this method of calculating acceleration? Explain. Does it agree with your prediction?

3.6. VELOCITY AND ACCELERATION FOR CHANGING MOTION

In the activities in the last section you looked at position and velocity vs. time graphs of the motion of a cart that moved at a constant velocity. The data for these graphs were collected using a motion sensor. Your goal in this section is to use a motion sensor to track the motion of a cart as its velocity changes at a constant rate. For the activities in this section you will need:

- 1 computer-based laboratory system
- 1 ultrasonic motion sensor
- 1 motion software
- 1 low-friction dynamics cart
- 1 ramp, 2 m (or smooth, level surface)
- 1 fan assembly
- 4 AA batteries
- 2 AA aluminum dummy cells

Recommended group size:	3	Interactive demo OK?:	Y

Speeding Up at a Moderate Rate

In the next activity you will look at velocity and acceleration graphs of the motion of a cart when its velocity is increasing at a moderate rate. You will be able to see how these two representations of the motion are related to each other when the cart is speeding up.

In order to get your cart speeding up smoothly you can use a propeller driven by an electric motor to accelerate the cart. Your task in the next activity is to create nice smooth graphs of position and velocity vs. time for a cart driven by a fan assembly. You should set up the cart, ramp, fan attachment, and motion sensor as shown in the following diagram.

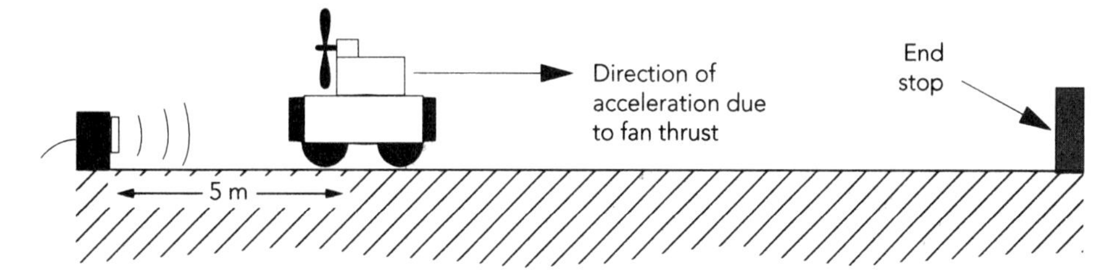

Fig. 3.14. Diagram of setup of cart, fan assembly, motion sensor, and ramp with an end stop mounted on it. Note that the fan blade does not stick out beyond the end to the cart.

1. Use two batteries and two dummy cells in the fan assembly so that the cart does not speed up too fast.
2. Always catch the fan at the end of a run before it crashes!
3. To keep the motion sensor from collecting bad data from the rotation of the propeller, be sure that the fan blade *does not extend beyond the front end of the cart.*
4. Set up the motion software to display *Position* from 0 to 2.0 m and *Velocity* from –1.0 to 1.0 m/sec for a total time of 3.0 sec as shown in the following graph diagram.
5. If necessary, rescale the graphs so that the traces fill the screen.
6. Practice several times before completing the activities described below.

3.6.1. Activity: Graphs Depicting Speeding Up

a. Predict the shape of your position vs. time and velocity vs. time graphs of your fan and cart as they move away from the sensor and speed up. Sketch the graphs neatly on the following axes using *dashed* lines.

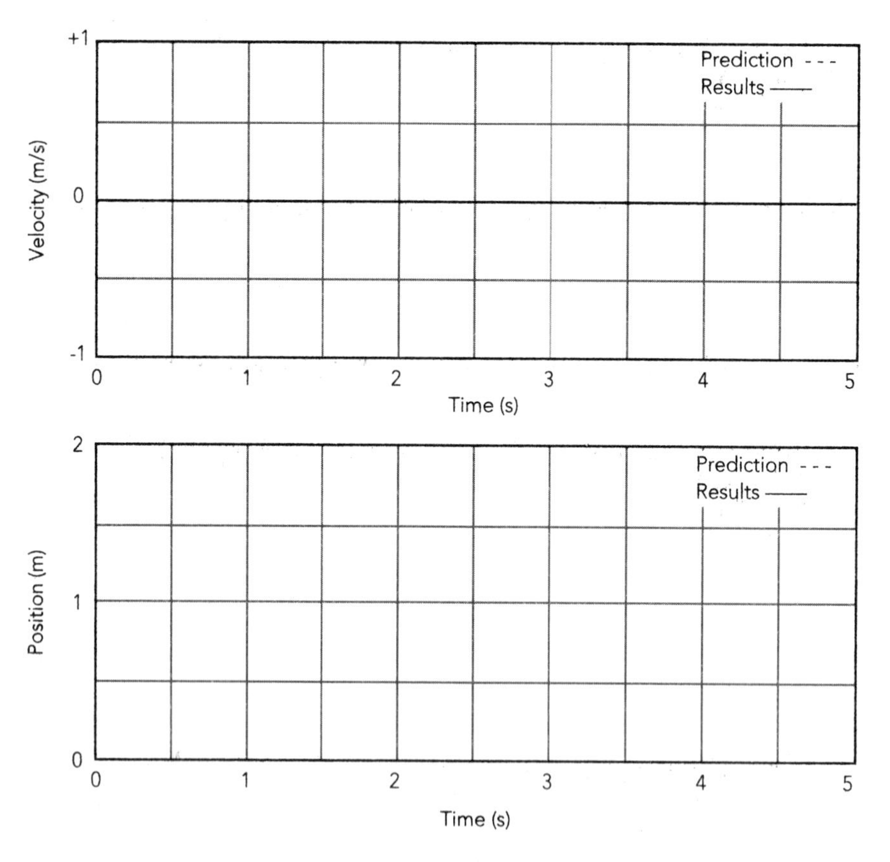

b. Create position vs. time and velocity vs. time graphs of your fan cart as it moves away from the sensor and speeds up. Sketch the graphs neatly on the preceding axes with a solid line.

c. Save this set of data and graphs for future use and for comparison with graphs you will make in the next activity of a more rapid change in motion.

d. How does your position graph differ from the position graphs for steady (constant velocity) motion?

e. What feature of your velocity graph signifies that the motion was *away* from the sensor?

f. What feature of your velocity graph signifies that the cart was *speeding up*? How would a graph of motion with a constant velocity differ?

g. Change the *Position* display to *Acceleration.* Adjust the acceleration scale so that your graph fills the axes. Sketch your graph on the acceleration axes that follow.

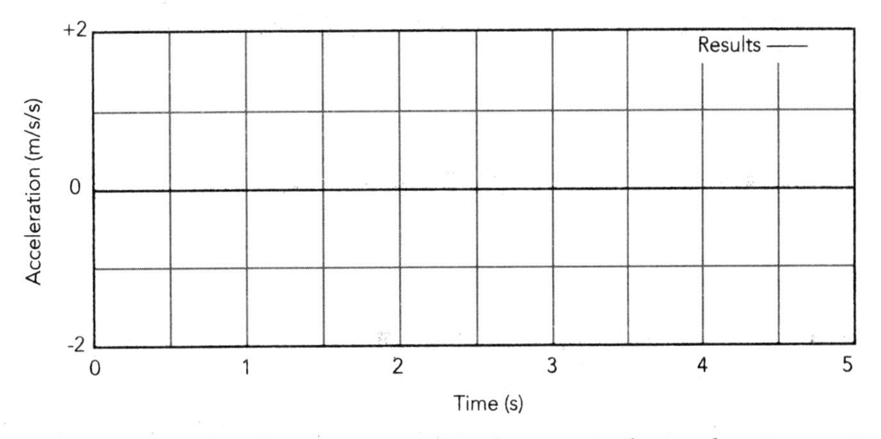

h. During the time that the cart is speeding up, is the acceleration positive or negative? How does *speeding up* while moving *away* from the sensor result in this sign of acceleration? **Hint:** Remember that acceleration is the *rate of change* of velocity. Look at how the velocity is changing.

i. How does the velocity vary in time as the cart speeds up? Does it increase at a steady rate or in some other way?

j. How does the acceleration vary in time as the cart speeds up? Is this what you expect based on the velocity graph? Explain.

Speeding Up at a Faster Rate

Suppose that you accelerate your cart at a faster rate by putting four batteries instead of two in it? How would your velocity and acceleration graphs change?

Fig. 3.15. Diagram of cart and fan assembly driven by four batteries for an increased change in motion.

Note: You should arrange to collect data in the following activity in such a way that its graph can be compared to the graph you obtained in Activity 3.6.3 directly on the computer screen.

3.6.2. Activity: Observing a Larger Acceleration

a. Resketch the "two-battery" velocity and acceleration graphs you found in Activity 3.6.1 once more using the following axes. Then predict how the four battery results might look using a dashed line.

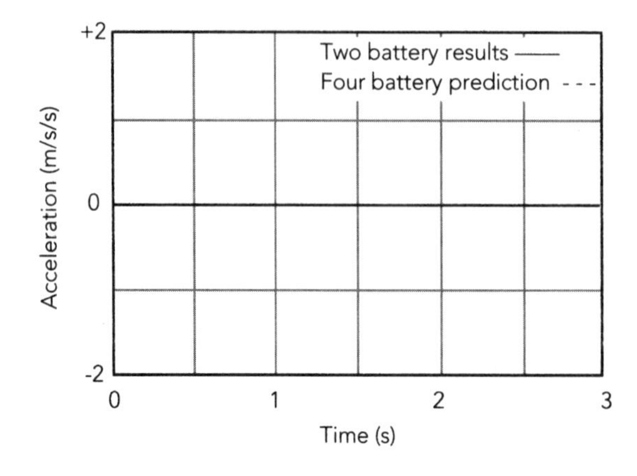

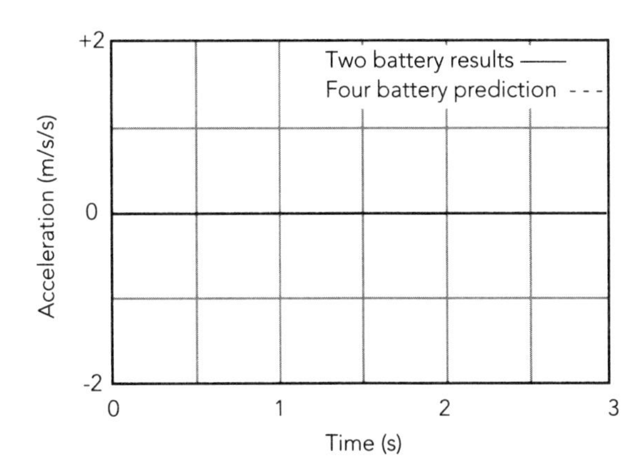

b. Test your predictions by accelerating the cart with four AA cells in the fan-assembly battery compartment. Repeat, if necessary, to get nice graphs and then sketch the results in the axes that follow. *Use the same scale as you did for the sketch of the two-battery graph.*

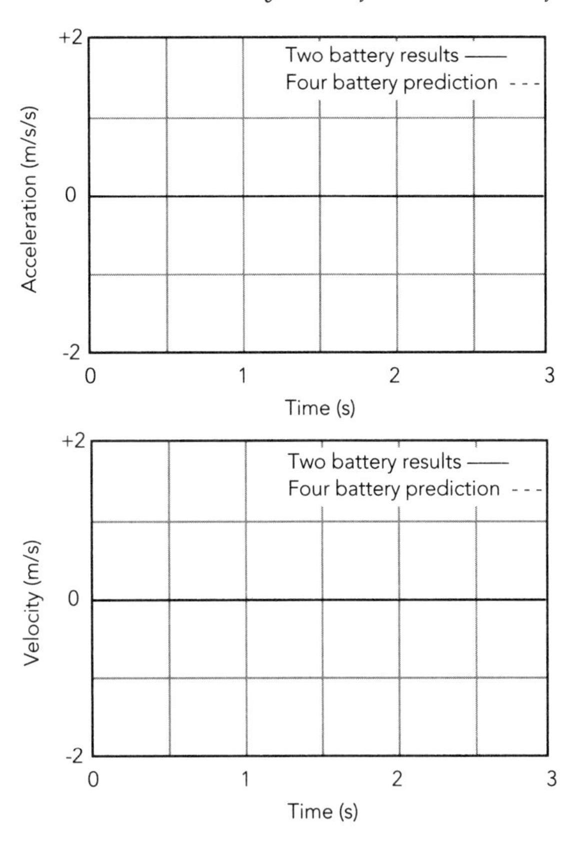

d. Did the general shapes of your velocity and acceleration graphs agree with your predictions? How is the greater magnitude (size) of acceleration represented on a velocity vs. time graph?

e. How is the greater magnitude (size) of acceleration represented on an acceleration vs. time graph?

Note: Please save your data for future use.

MATHEMATICAL DEFINITIONS OF VELOCITY AND ACCELERATION

3.7. CALCULATING VELOCITY AND ACCELERATION FROM POSITION DATA

By now you should have a good intuitive feeling for how to describe motions in terms of the changes in position, velocity, and acceleration that an object might undergo. You can tell people about these motions, graph them, or draw a picture of an object at various times with velocity vectors showing its relative speed and direction. However, the motion software was written to calculate velocity and acceleration from position data for you. In this session you will learn how average and instantaneous velocity and acceleration are defined mathematically, and you will gain experience with using these definitions to calculate these quantities from position data. You will start by generating your own position data from a photograph of the motion of a low-friction cart propelled by a fan assembly. This data can then be used in subsequent calculations of velocity and acceleration.

Measuring Position as a Function of Time

In order to understand more about how the motion software actually translates measurements into one-dimensional velocities and accelerations, it is helpful to make your own length and time measurements for a cart system.

Consider the type of constantly accelerated cart motion that you studied in the last two sessions. Suppose that instead of a motion sensor you have a video camera off to one side so you can film the location of the cart 30 times each second. (This is the rate at which a standard video camera records frames.) By displaying frames at regular time intervals it is possible to view the position of the cart on each frame as shown in Figure 3.16.

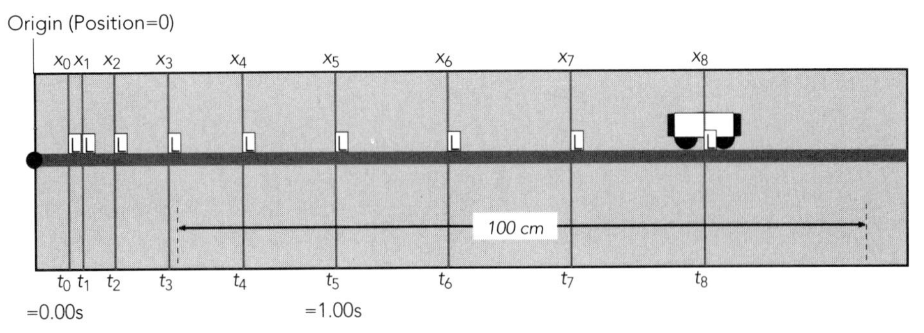

Fig. 3.16. A scale diagram of the position of an accelerating cart at eight equally spaced time intervals. The cart actually moved a distance of just less than 1 meter. A meter stick was placed in the plane of the cart motion for scale. Every sixth frame was displayed in the cart movie, so that five frames were recorded each second. *At each time the center of the cart is located in the upper left corner of the rectangle with a number 1 in it.*

For the next activity you will need:

- 1 ruler (with a metric scale)
- 1 demo cart
- 1 ramp

Recommended group size:	2	Interactive demo OK?:	N

Optional: If your class wants to record, digitize, and analyze a video frame sequence of a cart motion, the following equipment is needed:

- 1 low-friction cart
- 1 ramp, 2 m
- 1 computer-based laboratory system
- 1 video camera (with cables) and video board (in computer) or 1 digital movie of an accelerating cart
- 1 VideoPoint software

Recommended group size:	All	Interactive demo OK?:	Y

3.7.1. Activity: Position vs. Time from a Cart Video

a. Start the measuring process by recording key scaling factors for calibration. How much time, Δt, has elapsed between frame 0 and frame 1, between frame 1 and frame 2, etc. What is the calibration factor (i.e., how many real meters are represented by each centimeter or pixel in the diagram)?

D*t* = ________ *s*

m/cm = ________

or

m/pixel = ________

b. Use a ruler to measure the cart's distance from the origin (i.e., its position) in cm at each of the times 0.00 s, 0.20 s, etc. and fill in columns 1 and 2 in the table that follows.

Optional alternative: If a computer-based video of this cart motion is available, determine the cart's distance from the origin in pixels (1 pixel = 1 picture element) using video analysis software.

Frame #	Position	cm (or pixels) from origin in diagram 1 $x(\quad)$	Elapsed time (s) 2 t(s)	Actual distance from origin (m) 3 x(m)
0	x_0		0.000	
			0.100	
1	x_1		0.200	
			0.300	
2	x_2		0.400	
			0.500	
3	x_3		0.600	
			0.700	
4	x_4		0.800	
			0.900	
5	x_5		1.000	
			1.100	
6	x_6		1.200	
			1.300	
7	x_7		1.400	
			1.500	
8	x_8		1.600	

c. Use the scaling factor between "diagram centimeters or pixels" and real meters to calculate the position in meters of the cart. You'll want to use a spreadsheet for this. Place the results in column 3.

3.8. DEFINING AVERAGE VELOCITY MATHEMATICALLY

By considering the work you did with the motion sensor and with the measurements you just performed in Activity 3.7.1, you should be able to define average velocity along a line in words or even mathematically. *Remember that velocity is a rate of change of position divided by the time interval over which the change occurred.*

Note: Mathematically, change is defined as the difference between the final value of something minus the initial value of something.

$$\text{Change} \equiv (\text{Final Value}) - (\text{Initial Value})$$

3.8.1. Activity: Defining Velocity in One Dimension

a. Describe in *words* as accurately as possible what the word "velocity" means by drawing on your experience with studying velocity graphs of motion. **Hint:** How can you tell from the graph the direction an object moves? How can you tell how fast it is moving?

b. Suppose that you have a long tape measure and a timer to keep track of a cart or your partner who is moving irregularly along a line. For the purposes of this analysis, assume that the object of interest is a mere point mass. Describe what you would need to measure and how you would use these measurements to calculate velocity at a given moment in time.

c. Can you put this description in mathematical terms? Denote the average velocity with the symbol $\langle v \rangle$. Suppose the distance from the origin (where the motion sensor was when it was being used) to your partner is x_1 at a time t_1 just before the moment of interest and that the distance changes to x_2 at a later time t_2, which is just after the moment of interest. Write the equation you would use to calculate the average velocity, $\langle v \rangle$, as a function of x_1, x_2, t_1, and t_2. What happens to the sign of $\langle v \rangle$ when x_1 is greater than x_2?

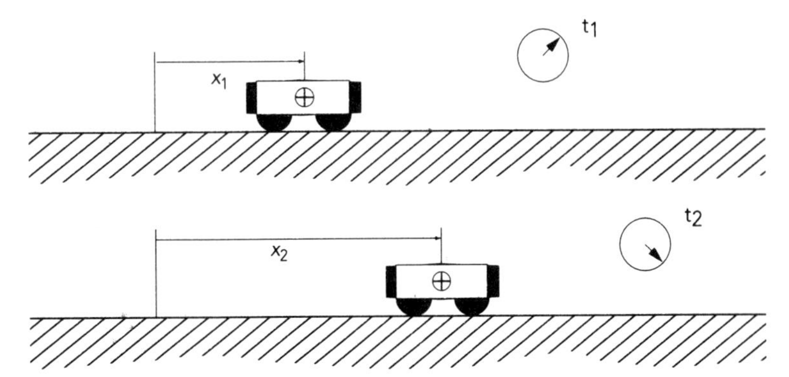

Fig. 3.17. Rolling on a track.

d. Use the mathematical definition in part c. to calculate each of the average velocities of the cart motion described in Activity 3.7.1 and fill in columns 3 and 4 in the table that follows. You will want to use a spreadsheet for this calculation. **Important note:** $t_2 - t_1$ represents a time interval, Δt, between two measurements of position and is not necessarily the total time that has elapsed since a clock started.

e. Show at least one sample calculation in the space below.

Elapsed time (s) 2 t(s)	Actual distance from origin (m) 3 x(m)	Average velocity (m/s) 4 (m/s)	Average acceleration(m/s) 5 (m/s/s)
0.000			
0.100			
0.200			
0.300			
0.400			
0.500			
0.600			
0.700			
0.800			
0.900			
1.000			
1.100			
1.200			
1.300			
1.400			
1.500			
1.600			

3.9. DEFINING AVERAGE ACCELERATION MATHEMATICALLY

By considering the work you did with the motion sensor, you should be able to define average acceleration in one dimension mathematically. It is similar to the mathematical definition of average velocity you developed in Activity 3.8.1. All the circumstances in which accelerations are positive and negative are described by the equation that defines them.

A pirate is about to throw you overboard. As a consequence you start to walk more and more slowly along a plank. You are accelerating! Why?

3.9.1. Activity: Defining Average Acceleration

a. Describe in *words* as accurately as possible what the word "acceleration" means by drawing on your experience with studying velocity *and* acceleration graphs of motion.

b. Suppose your *average* velocity is $\langle v_1 \rangle$ at a time t_1 and that the *average* velocity changes to $\langle v_2 \rangle$ at a later time t_2. Write an equation for the average acceleration in the space below.

Fig. 3.18. Walking the plank slower and slower!

c. Use the mathematical definition in part a. to calculate the average acceleration of the cart motion depicted in Activity 3.7.1 and fill in column 5 in Activity 3.8.1c above for each time interval (i.e., 0.00 to 0.20 s, 0.20 s to 0.40 s, etc.). You will want to use a spreadsheet for this calculation. Show at least one sample calculation in the space below.

d. Suppose you are walking away from a motion sensor as shown above. How does your rate of your walking change if $\langle v_1 \rangle$ is greater than $\langle v_2 \rangle$? Is your acceleration positive or negative? Use the mathematical equation in part a. to explain your answer.

e. Suppose you are walking toward a motion sensor. How is your speed (i.e., magnitude of velocity) changing for $\langle v_1 \rangle > \langle v_2 \rangle$? Is your acceleration positive or negative? *Be very careful with your mathematics on this one. It's tricky!*

f. Is the sign of $\langle a \rangle$ always the same as the sign of the velocities? Why or why not? **Hint:** What happens when you walk the plank?

3.10. ACCELERATION AS THE SLOPE OF A VELOCITY GRAPH

Just as velocity is the rate of change of position, acceleration is the rate of change of velocity. How do we find the acceleration of an object at a single instant (i.e., during a time interval that is too small to measure directly)? Since acceleration is the rate of change of velocity, the acceleration of an object is given by the slope of a smooth curve drawn through its velocity vs. time graph. (See Appendix D for tips on fitting curves to data with uncertainties.)

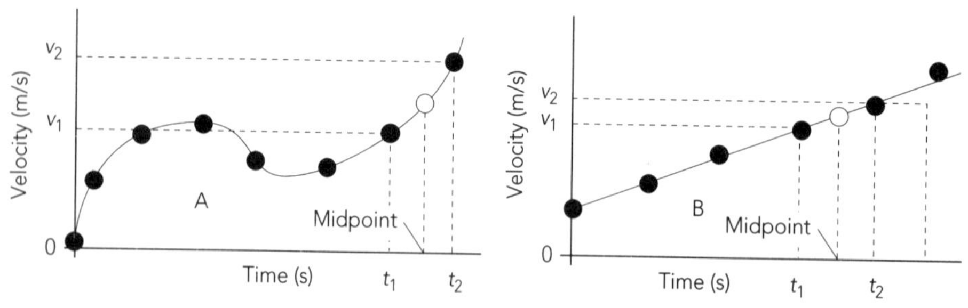

Fig. 3.19. Graphs of velocity vs. time based on data with little uncertainty. A smooth curve can be sketched near the data points. The instantaneous acceleration can be approximated as the slope of the curve that fits the velocity data. Graph A shows non-uniform acceleration; graph B shows uniform acceleration.

Let's apply this graphical analysis approach to the task of finding the instantaneous acceleration for the cart motion described in Activity 3.7.1.

3.10.1. Activity: Accelerations from the Cart Data

a. Refer to the data that you analyzed and recorded in Activity 3.8.1c. Create a graph of the average velocity, $\langle v \rangle$, as a function of time and affix it in the space below. **Note:** If it is available, you'll want to use an *Excel 4.0 Modeling Template* to enter data and plot this graph.

b. Find the slope of a line that seems to fit smoothly through most of the data points. The slope represents the acceleration of the cart. Report its value in the space below. **Note:** If you have an *Excel Modeling Worksheet* available, you should use mathematical modeling to find the best line that fits the data. (See Appendix E and the Excel Modeling Tutorial Worksheet for hints on how to do mathematical modeling.)

c. How does the value for acceleration that you determined from the slope compare to those you determined earlier from the average accelerations at the mid-point of each time interval between average velocity values. In other words, how does the slope compare with the averages reported in column 5 of the table in Activity 3.8.1c? Find the average and standard deviation of the elements in column 5 of that table.

$\langle a \rangle$ (m/s^2) =

σ_{sd} (m/s^2) =

d. Does the acceleration determined by the slope lie within one standard deviation of the average of the average accelerations?

You should have just verified that finding the average velocities and then average accelerations for a uniformly accelerated cart leads to the same average acceleration as finding the slope of the velocity vs. time graph. This method of determining acceleration by finding the slope (or, as they say in calculus, the derivative) of the velocity vs. time function is a method you will use many times in physics. It is one of the easiest and most accurate techniques for determining accelerations from measurements.

3.11. DETERMINING INSTANTANEOUS VELOCITIES AND ACCELERATIONS FROM SIMPLE EQUATIONS

Instantaneous velocity is defined as the time derivative of the function that describes how position changes in time. Similarly, instantaneous acceleration is defined as the time derivative of the function that describes how the instantaneous velocity varies in time. Thus, for an object moving in one dimension

$$v \equiv \frac{dx}{dt} \quad \text{and} \quad a \equiv \frac{dv}{dt}$$

Note: The triple bar symbol ($\equiv$) is stronger than an equality. It means "defined as."

In the special case in which the variation of a position or velocity with time can be represented by a power of time (bt^n), the derivative can be taken quite easily. The instantaneous velocity is given by

$$v \equiv \frac{dx}{dt} = nbt^{(n-1)}$$

For example, if $x = 5t^3$, then

$$v \equiv \frac{dx}{dt} = 5(3)t^2 = 15\ t^2$$

The acceleration can be found by taking the time derivative of the velocity function, v.

3.11.1. Activity: Determining *v* and *a* by Differentiation

a. Suppose $x = 4t^2$ m. Find a general expression for vx as a function of time. What is v at $t = 0$ s? At $t = 2$ s? (Don't forget units!)

$$v \equiv \frac{dx}{dt} =$$

b. Use the general expression for v_x you found in part a to find a general expression for the x-component of acceleration, a_x, as a function of time. What is the value of a at $t = 0$ s? At $t = 2$ s? Is the acceleration constant or does it change in time?

$$a \equiv \frac{dx}{dt} =$$

c. Suppose $x = 3t^4$. Find a general expression for v_x as a function of time. What is the value of v_x at $t = 0$ s? At $t = 2$ s?

d. Use the general expression for v_x you found in part c to find a general expression for the x-component of acceleration, a_x, as a function of time. What is the value of a_x at $t = 0$ s? At $t = 1$ s? Is the acceleration constant or does it change in time?

e. If the derivative of the sum of two functions is equal to the sum of the derivatives of those two functions, what is the instantaneous velocity of an object whose position is given by $x = 4t^2 + 3t^4$? **Hint:** Note that this is the sum of the functions in parts a. and c.

Name ______________________ Section __________ Date __________

UNIT 4: ONE-DIMENSIONAL MOTION II

A MATHEMATICAL DESCRIPTION OF CONSTANT ACCELERATION

Whether we think of a model as a scaled down (or scaled up) version of something or simply as an idealization, a model can only embody some aspects of the "real thing." For example, above is a small papier-mâché replica of a bird native to the Kashmiri Valley; it is handmade by local artisans and in a time-honored tradition is painted, sanded, and coated with many layers of lacquer. Although in this scale model the small bird lacks feet, a heart, and other organs, it does tell us about the shape of the bird and its exterior markings. There are no imperfections on the bird, so the model represents an idealization of a real one's exterior proportions. Also, this model is more compact because it takes up less space than the original; the bird is smaller than a real one would be.

Physicists can often describe the location of an object as a function of time with mathematical equations. Equations are smaller and more compact than actual motions, and they represent idealizations of motions because their graphs have perfect shapes. Graphs of real data can be messy. Physicists think of equations as elegant and compact mathematical models of important characteristics of "real situations." In this unit, you will learn how to create mathematical models of objects moving with constant velocity and constant acceleration.

UNIT 4: ONE-DIMENSIONAL MOTION II

A MATHEMATICAL DESCRIPTION OF CONSTANT ACCELERATION

At the point where a speeding driver is caught by a cop, the cop comes up to the speeder and says, "You were going 60 miles an hour!" The driver says, "That's impossible, I was traveling for only 7 minutes. It is ridiculous—how can I go 60 miles an hour when I wasn't going an hour?"

Richard Feynman, adapted from a joke in *The Feynman Lectures on Physics*, V. 1

OBJECTIVES

1. To recognize the pattern of position vs. time, velocity vs. time, and acceleration vs. time for the motion of objects that speed up, slow down, or turn around at a constant rate and to recognize this type of motion as constantly accelerated motion.

2. To use *mathematical modeling* techniques to determine equations describing one-dimensional motion with constant acceleration.

3. To derive and check the validity of some of the standard *kinematic equations* used to describe the motion of objects undergoing constant acceleration.

4. To learn to use the *kinematic equations* you derived to describe the motion of objects moving with constant acceleration.

4.1. OVERVIEW

In this unit you will continue the study of the motion of a low-friction cart propelled by a fan. In the first set of activities you will use a computer-based laboratory system with motion software to observe acceleration as the cart speeds up, slows down, and turns around.

The second set of activities involves the use a spreadsheet for *mathematical modeling* to find idealized equations that describe how the position of the cart changes in time when it undergoes a constant acceleration.

In the last series of activities you will work with the famous kinematic equations used to describe the motion of objects that have constant accelerations. These *kinematic equations* can be deduced logically based on the definitions of instantaneous velocity and acceleration and accepted rules of mathematics. You will be asked to derive these equations, and finally you will practice using them to describe the motions of various objects. The activities in this unit will complete your quest to learn to represent one-dimensional motion in *words, graphs, pictures,* and *equations.*

SPEEDING UP, SLOWING DOWN, AND TURNING AROUND

4.2. WORDS CAN BE AMBIGUOUS

In the past few sessions we have been using the term speed or magnitude of velocity to describe how fast or slow an object is moving. Speed can be calculated as the distance moved by the object divided by the time interval that elapsed as it moved that distance. For an object moving in a straight line, we can define the velocity of an object moving at a certain speed by choosing a coordinate system and putting a + or − sign in front of the speed to indicate what direction the object is moving along a coordinate axis.

In our study of the velocity and velocity changes of moving objects the careless use of words can be very confusing. Let's think about and discuss the meaning of some terms that people sometimes use to compare velocities such as greater than and less than, smaller and larger, faster and slower, and, finally, more positive and more negative.

What does the term deceleration mean? Which of the comparative terms make sense when talking about velocities and accelerations, and which are ambiguous? What do the words mean to mathematicians and scientists?

The Number Line: Greater Than and Less Than

When using pure numbers, the idea of *greater than* or *less than* makes sense. Mathematicians have devised a standard number line with positive numbers placed to the right and negative numbers placed to the left. They then define *greater* to mean more positive (further to the right) and *less* to mean further to the left on a number line. Sometimes people carelessly refer to numbers as being *larger* or *smaller* numbers in the same sense.

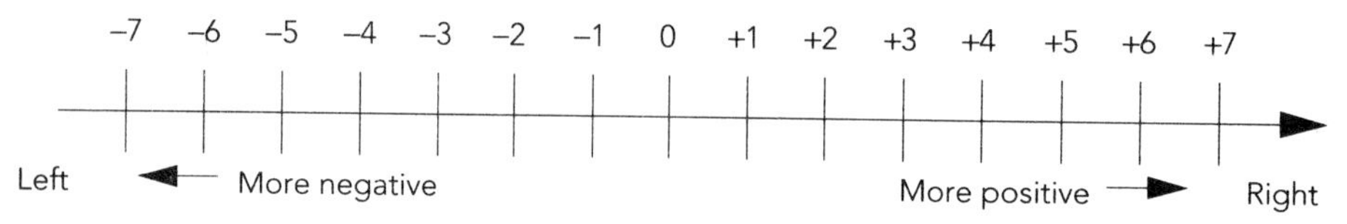

Fig. 4.1. A mathematical number line.

Describing Velocities

When describing velocities, the terms *larger* and *smaller* seem ambiguous. If the term larger is used without careful definition, we could mean something with a larger velocity is moving at a faster speed or we could perhaps mean the velocity is more positive. *Thus, in comparing velocities, the terms greater than and less than and larger than and smaller than should be avoided whenever possible.*

A less ambiguous way to compare velocities is to talk explicitly: (1) about an object having a *faster* or *slower* speed or magnitude of velocity, and (2) about an object moving in a *positive* or *negative* direction (relative to the chosen coordinate system).

Finding Acceleration Directions from Velocity Changes

Acceleration represents changes in velocities. It is important to recognize that acceleration, like velocity, behaves like a vector quantity. It has both magnitude and direction. The magnitude of the acceleration represents how much change there has been in the velocity and the sign of the acceleration represents the direction of the change in acceleration. The direction of change can be found using algebra as follows:

1. The acceleration is *positive* if the final velocity after a time interval, Δt, has elapsed is *either more positive* (*or less negative*) than the initial velocity;
2. The acceleration is negative if the final velocity after a time interval, Δt, has elapsed is more *negative* (*or less positive*) than the initial velocity.

Note: *More positive* means the same thing as *less negative* and vice versa.

This can be seen by using arrows to represent the magnitude and direction of the velocity vectors like those used in motion diagrams. Suppose an object has an initial velocity of +7.0 m/s (so it is moving to the right) and then after 1.5 seconds it has a final velocity of +4.0 m/s (so it is still moving to the right). This is illustrated in Figure 4.2.

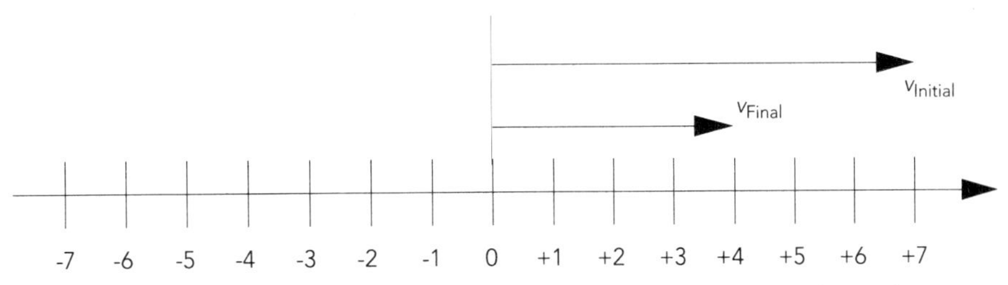

Fig. 4.2. A number line with vectors (i.e., directed arrows) showing an object slowing down while moving to the right.

To find the direction of the acceleration we can look at the change in velocity over the time period as having the same direction as the arrow representing the vector *difference* between the two velocities. This is shown in Figure 4.3.

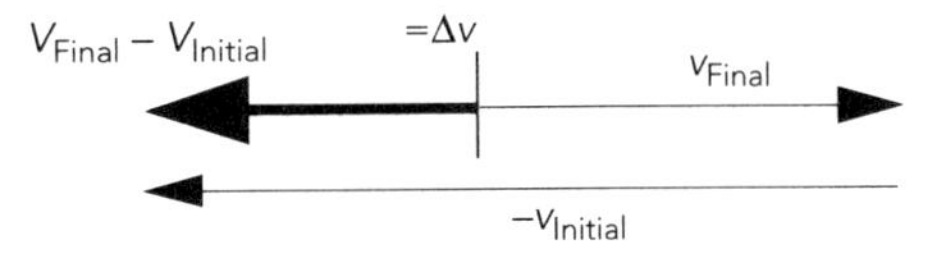

Fig. 4.3. The vector difference between an initial and final velocity.

Note that the vector representing the difference between the initial and final velocity is pointing to the left and is negative. This means the acceleration or direction of the change in the velocity is *negative* even though both velocities are *positive.* Since the magnitude of the change is 3.0 m/s and the time interval is 1.5 s, the average acceleration during the 3.0 s time interval in question is –2.0 m/s^2.

Describing Accelerations

In describing accelerations you should refer to the magnitude of acceleration as being large or small and specify whether it is positive or negative. In the activity that follows use the method of vector differences or algebra to determine the direction and magnitude of the acceleration for the following situations.

4.2.1. Activity: What Is the Sign of the Acceleration?

a. A car's velocity changes from +20 mi/hr to +16 mi/hr in 3 seconds. Is the average acceleration (change in velocity over that time interval) positive or negative? What is the magnitude of acceleration?

b. A runner's velocity changes from –12 m/s to –8 m/s in 1 second. Is the average acceleration (change in velocity over the time interval) positive or negative? What is the magnitude of acceleration?

c. At the beginning of a half-hour time period, a snail is moving at –1.0 cm/s; the snail then stops and turns around and starts heading back in the opposite direction at +0.2 cm/s. Is the average acceleration (change in velocity over the time interval) positive or negative? What is the magnitude of acceleration?

d. A car speeds up from –20 mi/hr to –30 mi/hr in 2 seconds. Is the average acceleration (change in velocity over the time interval) positive or negative? What is the magnitude of acceleration?

e. A donkey pulls a cart at +1 m/s and is still pulling at +1 m/s after an hour. Is the average acceleration (change in velocity over the time interval) positive or negative? What is the magnitude of acceleration?

4.3. ABOUT SLOWING DOWN AND SPEEDING UP

In the motions you studied in the last unit the velocity and acceleration vectors representing the motion of the cart both pointed in the same direction.

To get a better feeling for acceleration, it will be helpful for us to examine velocity vs. time and acceleration vs. time graphs for some more complicated motions of a cart on a ramp. Again you will use the motion sensor to observe the cart as it changes its velocity at a constant rate. Only this time the cart may be moving either toward or away from the sensor as it speeds up or slows down.

To complete the activities in this section you will need the same apparatus you used in Unit 3. This includes:

- 1 computer-based laboratory system
- 1 ultrasonic motion sensor
- 1 motion software
- 1 low-friction dynamics cart
- 1 ramp, 2 m (or smooth, level surface)
- 1 fan assembly
- 4 AA batteries
- 2 aluminum AA dummy cells

Recommended group size:	3	Interactive demo OK?:	N

Slowing Down

In this activity you will look at a cart (or toy car) moving along a ramp *and slowing down.* A car being brought to rest by the steady action of brakes is a good example of this type of motion. Later you will examine the motion of the cart *toward* the motion sensor and *speeding up.* In both cases, we are interested in the shapes of the velocity vs. time and acceleration vs. time graphs, as well as the vectors representing velocity and acceleration.

Let's start with the creation of velocity and acceleration graphs of when it is moving *away from* the motion sensor and *slowing down.* To do this activity, you should set up the cart, ramp, and motion sensor as shown below. If the cart has a friction pad, move it out of contact with the ramp so that the cart can move freely. Use the same two AA cells and two dummy cells as you used for the first *Speeding Up* activity in Unit 3.

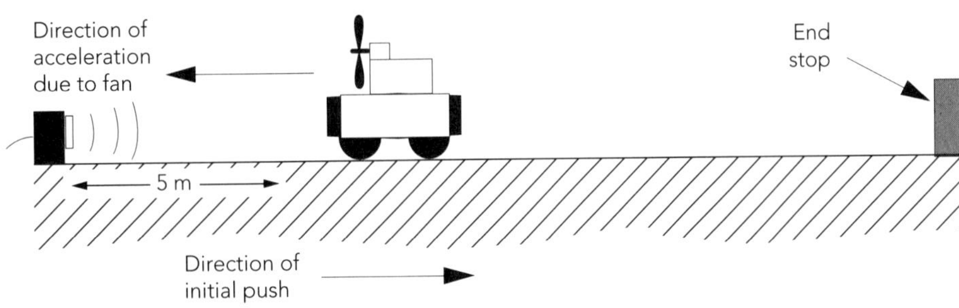

Fig. 4.4. Diagram of setup of cart, fan assembly, motion sensor, and ramp for observing the cart slowing down.

Now when you give the cart with its fan on and thrusting toward the motion sensor a push away from the motion sensor, it will slow down after it is released. In this activity you will examine the velocity and acceleration of this motion.

4.3.1. Activity: Graphs Depicting Slowing Down

a. If you give the cart a push away from the motion sensor and release it, will the acceleration be positive, negative, or zero (after it is released)? Think about the direction of change of the velocity (i.e., use your conclusions from Activity 4.2.1) to sketch your predictions for the velocity vs. time and acceleration vs. time graphs on the axes below using a dashed line.

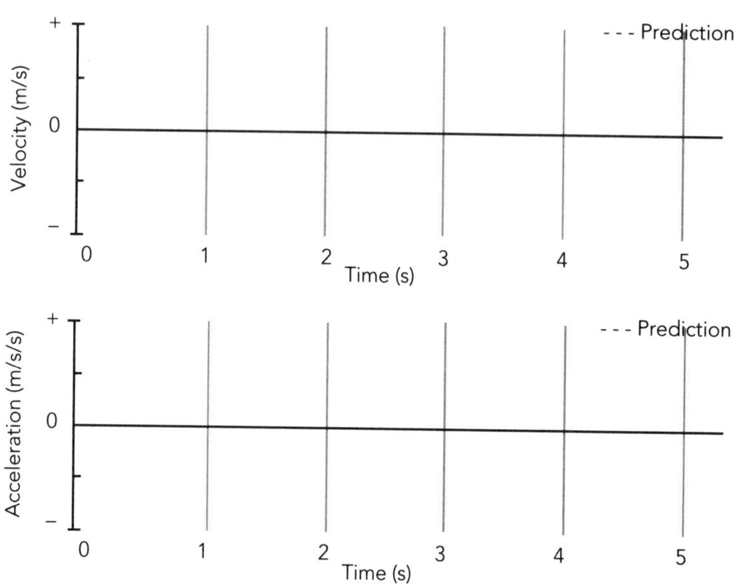

b. Test your predictions by opening the motion software and setting it up to display velocity and acceleration graphs like those shown in the following axes. Next locate the cart 0.5 m from the sensor, turn on its fan, and push the cart away from the sensor when it starts clicking. Graph velocity first. Catch the cart before it turns around. You may have to try a few times to get a good run.

Sketch the results on the axes that follow. Don't forget to change the scales if this will make your graphs clearer before making the sketch.

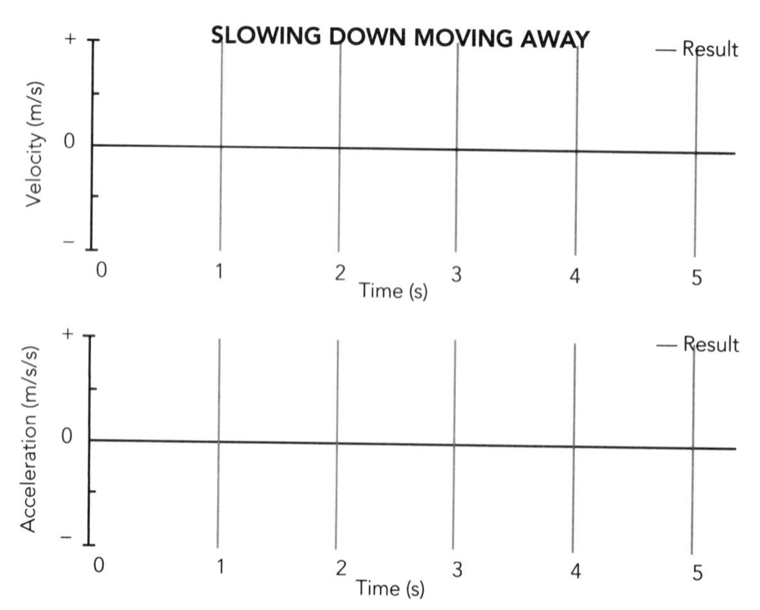

c. Label the graphs sketched on the above axes with—

- "A" at the spot where you started pushing.
- "B" at the spot where you stopped pushing.
- "C" at the spot where the cart stopped moving.

d. Did the shapes of your velocity and acceleration graphs agree with your predictions? How is the sign of the acceleration represented on a velocity vs. time graph?

e. How is the sign of the acceleration represented on an acceleration vs. time graph?

f. Is the sign of the acceleration what you predicted? How does *slowing down* while moving *away* from the sensor result in this sign of acceleration? **Hint:** Remember that acceleration is the *rate of change* of velocity. Look at how the velocity is changing.

g. Name and save this set of data and graphs on the computer for comparison with graphs you will make in the next activity and for future use.

Finding the Acceleration Direction for Slowing Down

Let's consider a motion diagram of a cart that is slowing down and use vector techniques to figure out the direction of the acceleration.

4.3.2. Activity: Vector Diagrams for Slowing Down

a. The following motion diagram shows the positions of the cart at equal time intervals. (This is like taking snapshots of the cart at equal time intervals.) At each indicated time, sketch a vector above the cart that might represent the velocity of the cart at that time while it is moving away from the motion sensor and slowing down.

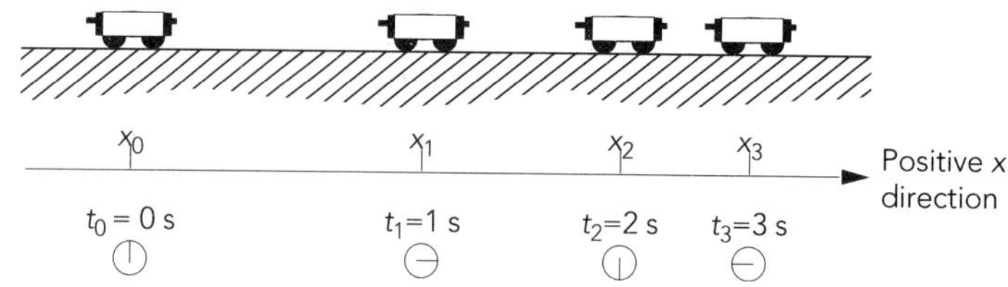

Fig. 4.5.

b. Show below how you would find the vector representing the change in velocity between the times 1 s and 2 s in the above diagram. Based on the direction of this vector and the direction of the positive x-axis, what is the sign of the acceleration?

c. Based on your observations in this activity and in the last session, state a general rule to predict the sign and direction of the acceleration if you know the sign of the velocity (i.e., the direction of motion) and whether the object is speeding up or slowing down.

Speeding Up Toward the Motion Sensor

Let's investigate another common situation. Suppose the cart is allowed to speed up when traveling toward the motion sensor. What will the direction of its acceleration be? Positive or negative?

4.3.3. Activity: Graphs Depicting Speeding Up

a. Try to predict the shapes of the velocity and acceleration graphs for speeding up toward the motion sensor. Think about the direction of change of the velocity (i.e., use your conclusions from Activity 3.6.2) to sketch your predictions for the velocity vs. time and acceleration vs. time graphs on the axes that follow.

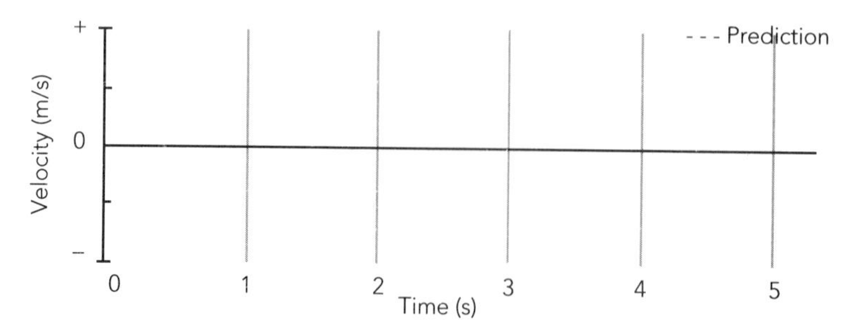

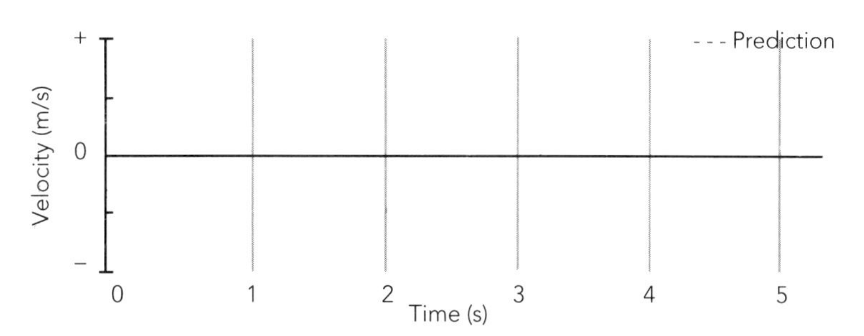

b. Test your predictions by turning on the fan and releasing the cart from rest at the end of the ramp after the motion sensor starts clicking. Catch the cart before it gets too close to the sensor.

Draw the results on the axes that follow. You may have to try a few times to get a good run. Don't forget to change the scales if this will make your graphs clearer before making the sketch.

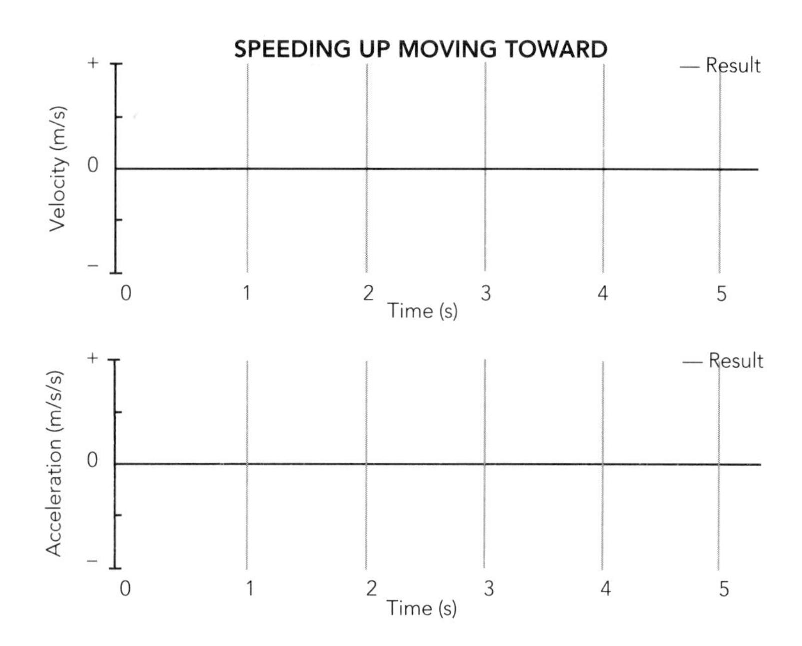

c. How does your velocity graph show that the cart was moving *toward* the sensor?

d. During the time that the cart was speeding up, is the acceleration positive or negative? Does this agree with your prediction? Explain how *speeding up* while moving *toward* the sensor results in this sign of acceleration. **Hint:** Think about how the velocity is changing.

Finding the Direction of Acceleration for Speeding Up

Let's consider a diagrammatic representation of a cart that is speeding up and use vector techniques to figure out the direction of its acceleration.

4.3.4. Activity: Vector Diagrams for Speeding Up

a. The following diagram shows the positions of the cart at equal time intervals. (This is like taking snapshots of the cart at equal time intervals.) At each indicated time, sketch a vector above the cart that might represent the velocity of the cart at that time while it is moving toward the motion sensor and speeding up.

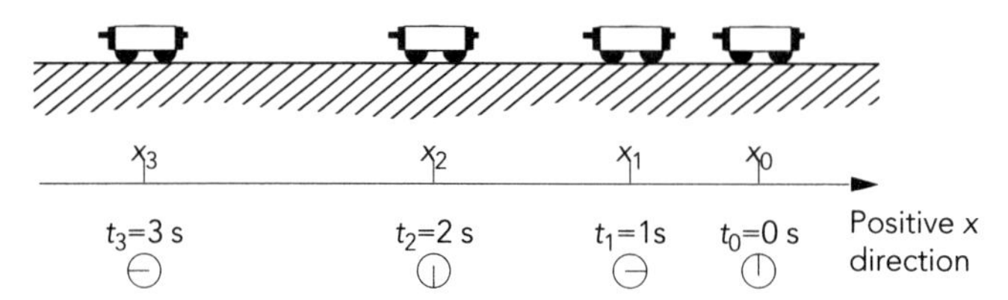

Fig. 4.6.

b. Show below how you would find the vector representing the change in velocity between the times 1 s and 2 s in the above diagram. Based on the direction of this vector and the direction of the positive x-axis, what is the sign of the acceleration?

Moving Toward the Sensor and Slowing Down

There is one more possible combination of velocity and acceleration for the cart, that of moving *toward* the sensor while *slowing down.*

4.3.5. Activity: Slowing Down Toward the Sensor

a. Use what you have learned so far to predict the direction and sign of the acceleration when the cart is slowing down as it moves toward the sensor. Explain why the acceleration should have this direction and this sign in terms of the velocity and how the velocity is changing.

b. The motion diagram in Fig. 4.7 shows the positions of the cart at equal time intervals for slowing down while moving toward the sensor. At each indicated time, sketch a vector above the cart that might represent the velocity of the cart at that time while it is moving toward the motion sensor and slowing down.

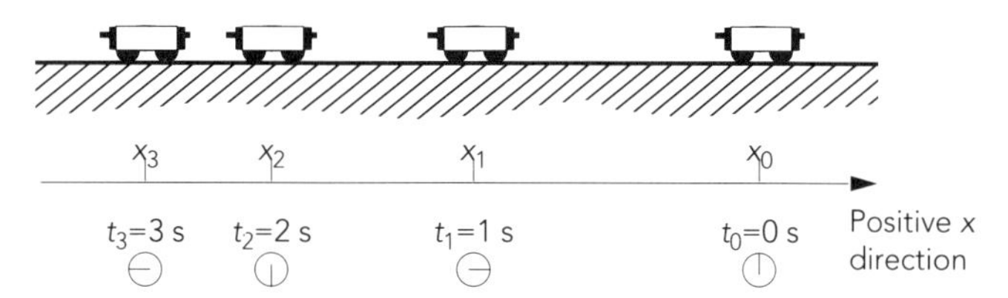

Fig. 4.7.

c. Show below how you would find the vector representing the change in velocity between the times 1 s and 2 s in the above diagram. Based on the direction of this vector and the direction of the positive x-axis, what is the sign of the acceleration? Does this agree with the prediction you made in part a?

4.4. ACCELERATION AND TURNING AROUND

In the next activity you will look at what happens when the cart slows down, turns around, and then speeds up. How is its velocity changing? What is its acceleration?

In order to complete the activities in this section you will need the same apparatus you used in the last section. This includes:

- 1 computer-based laboratory system
- 1 ultrasonic motion sensor
- 1 motion software
- 1 low-friction dynamics cart
- 1 ramp, 2 m (or smooth, level surface)
- 1 fan assembly
- 4 AA batteries
- 2 aluminum AA dummy cells

Recommended group size:	3	Interactive demo OK?:	N

The setup you can use is the same as for the study of a cart moving away from the detector and slowing down. Place two AA cells and two dummy cells in the fan unit.

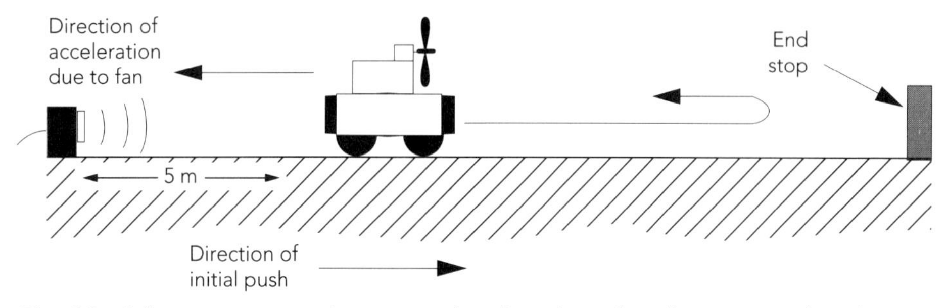

Fig. 4.8. A fan cart setup so that an initial push to the right will cause it to slow down and then turn around and speed up.

To practice this motion you should start the fan with its thrust toward the motion detector. Then give the cart a push *away* from the motion sensor. It moves toward the end of the ramp, slows down, reverses direction, and then moves back toward the sensor. Try it a couple of times before activating the motion sensor! *Be sure to stop the cart before it hits the motion sensor.*

4.4.1. Activity: Reversing Direction

a. For each part of the motion—away from the sensor, at the turning point, and toward the sensor—predict in the table that follows whether the velocity will be positive, zero, or negative. Also indicate whether the acceleration is positive, zero, or negative.

	Moving away	Turning around	Moving toward
Velocity			
Acceleration			

b. Sketch the predicted shapes of the velocity and acceleration graphs of this entire motion on the following axes.

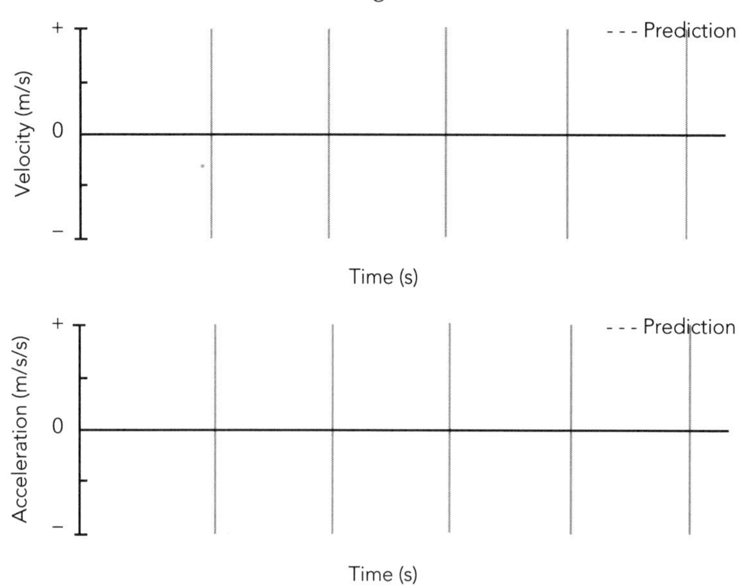

c. To test your predictions open the motion software and set up the velocity and acceleration axes as shown in the axes below. Use procedures that are similar to the ones you used in the slowing down and speeding up activities. *You may have to try a few times to get a good run.* Don't forget to change the scales if this will make your graphs clearer. When you get a good run, sketch both graphs on the axes.

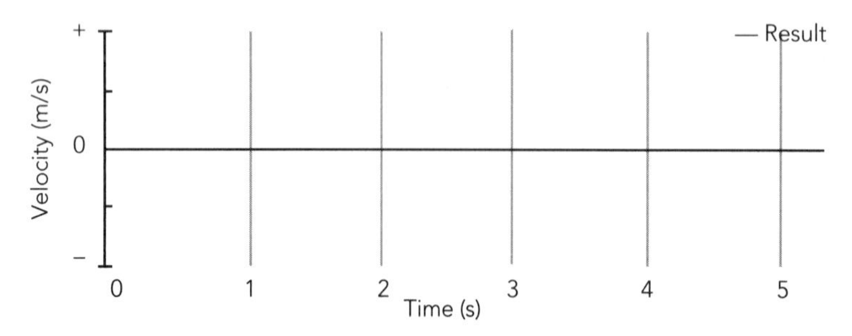

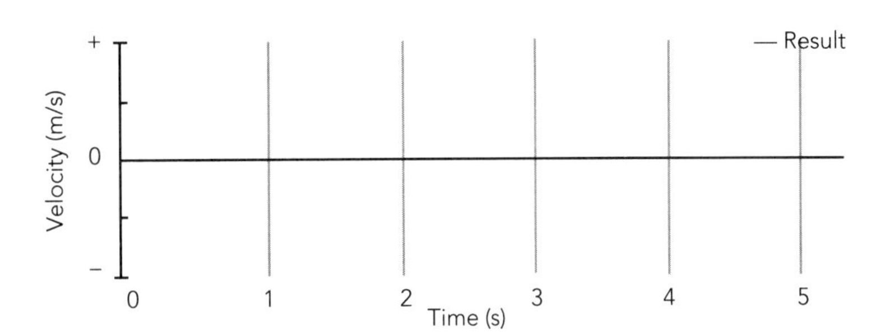

d. Label *both* graphs with—

- "A" where the cart started being pushed.
- "B" where the push ended (where your hand left the cart).
- "C" where the cart reached the end of the ramp (and is about to reverse direction).
- "D" where the you stopped the cart.

e. Did the cart have a zero velocity? Look at the velocity graph. Does this agree with your prediction? How much time did it spend at zero velocity before it started back toward the sensor? Explain.

f. According to your acceleration graph, what is the acceleration at the instant the cart is turning around? Is it positive, negative, or zero? Does this agree with your prediction?

g. Explain the observed sign of the acceleration near the end of the ramp. **Hint:** Remember that acceleration is the *rate of change* of velocity. When the cart is at rest at the end of the ramp, what will its velocity be in the next instant? Will it be positive or negative?

h. Set your motion software to display a position vs. time graph of your cart slowing down and then reversing direction. Use the space below to sketch your graph.

i. What is the *shape* of the position vs. time graph? The graph should look like one of the functions you studied in mathematics. If so, what type of function does it look like?

j. Name and save this set of data for use in the upcoming mathematical modeling activity.

Deceleration Is a Bad Word!

Often people refer to deceleration when an object is slowing down. Others might carelessly assume that deceleration means the object has a negative acceleration. You now know that it is possible for an object with either a negative or positive acceleration to be slowing down, turning around, and then speeding up. *Thus, it is best to avoid using the word deceleration even if it has more than four letters!*

Tossing a Ball

Suppose you throw a ball up into the air. It moves upward, reaches its highest point, and then moves back down toward your hand. What can you say about the directions of its velocity and acceleration at various points?

4.4.2. Activity: The Rise and Fall of a Ball

a. Consider the ball toss carefully. Assume that *upward* is the positive direction. Indicate in the table that follows whether the velocity is positive, zero, or negative during each of the three parts of the motion. Also indicate if the acceleration is positive, zero, or negative. **Hint:** Remember that to find the acceleration, you must look at the change in velocity.

	Moving up after release	At highest point	Moving down
Velocity			
Acceleration			

b. In what ways is the motion of the ball similar to the motion of the cart that you just observed?

Fig. 4.9.

c. Are there any parts of this up, turn around, down motion where you would say that the ball is "decelerating?" Is the term ambiguous?

MATHEMATICAL MODELING OF MOTION

4.5. USING EQUATIONS TO DESCRIBE CONSTANT ACCELERATION

It is cumbersome to use words and diagrams to convey much detailed quantitative information. Equations are inherently quantitative. Quantitative information about velocity and acceleration is contained in a graph of position vs. time. For this reason, graphs and equations are closely related. When an object is moving with either a constant velocity or a constant acceleration the equations used to describe simple motions can be graphed. Conversely, one can guess an equation and see if it fits a graph of motion based on actual data.

Fig. 4.10. An average person vs. a physics student thinking of a graph. The key element of this figure is a graph representing a motion with constant acceleration.

Fig. 4.11. An average person vs. a physics student thinking of an equation. The key element of this figure is an equation representing a motion with constant acceleration.

4.6. PARABOLAS AND ACCELERATED MOTION GRAPHS

In the next few activities we will explore the parabolic nature of the position vs. time graphs of a cart undergoing a constant acceleration. Then you will do some mathematical modeling on some of your old data. In particular, you will practice finding the appropriate equations to describe the position vs. time data you obtained using the motion detector and with the video analysis measurements.

To do next activity you will need the following equipment.

- 1 microcomputer
- 1 spreadsheet software
- 1 parabola worksheet (optional)

Recommended group size:	2	Interactive demo OK?:	N

Parabolas

In the activity on turning around you should have noticed that your graph of position vs. time looked like a parabola. In most basic mathematics courses the parabola is studied extensively as a simple example of a polynomial equation. Formally, a parabola is a second-order polynomial because the highest power of x is the second power. Mathematicians usually give the equation for the parabola as follows:

$$y = c_2x^2 + c_1x + c_0 \tag{4.1}$$

where c_2, c_1, and c_0 are constants. Although all parabolas have the same basic shape, some are skinny, some are fat, some start one place on a coordinate axis and some start another place. Some parabolas are even upside down! The magnitudes of the constants and whether they are positive or negative determine how a given parabola looks. In the next few activities you are going to use spreadsheet calculations and graphing to demonstrate to yourself how the values of each of the constants affect the shape of a parabola.

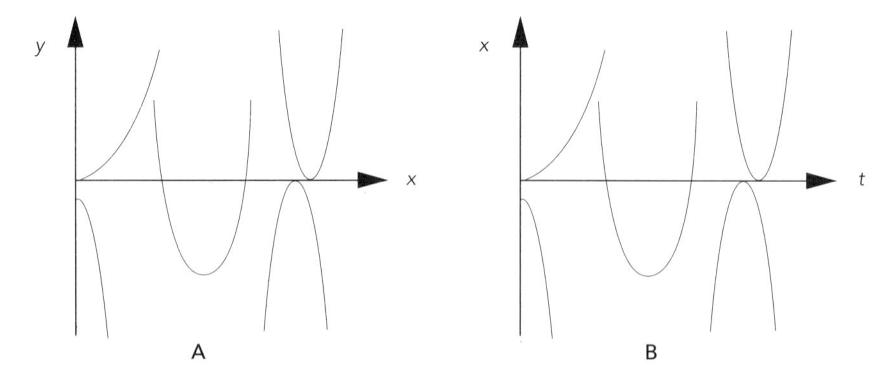

Fig. 4.12. Some skinny, fat, and upside down parabolas in different locations on a graph (A) that uses typical mathematics notation of y vs. x, and (B) that uses typical physics notation of x vs. t.

To do the following activity you should either use a preconfigured spreadsheet file (if it is available) or configure your own spreadsheet file as follows:

1. Set up a spreadsheet with initial values of c_0, c_1, and c_2 on it and columns for your x and y data as shown in the figure that follows—that is, for $c_0 = 1$, $c_1 = 1$, and $c_2 = 2$.
2. Enter Equation 4.1 as the first entry in the y-column. Be sure to use absolute references when referring to the constants.
3. Copy the equation down through all 20 rows in the y-column
4. Create a graph of your y vs. x values so you can examine the shape of the graph for different values of the constants.
5. Set the y-axis of the graph to remain fixed at a maximum value of $+800$ and a minimum value of -800. *If you don't do this, the graph scale will change as you try different values for the constants and it will be very difficult to figure out what changing the constants does to the shape of the graphs.*

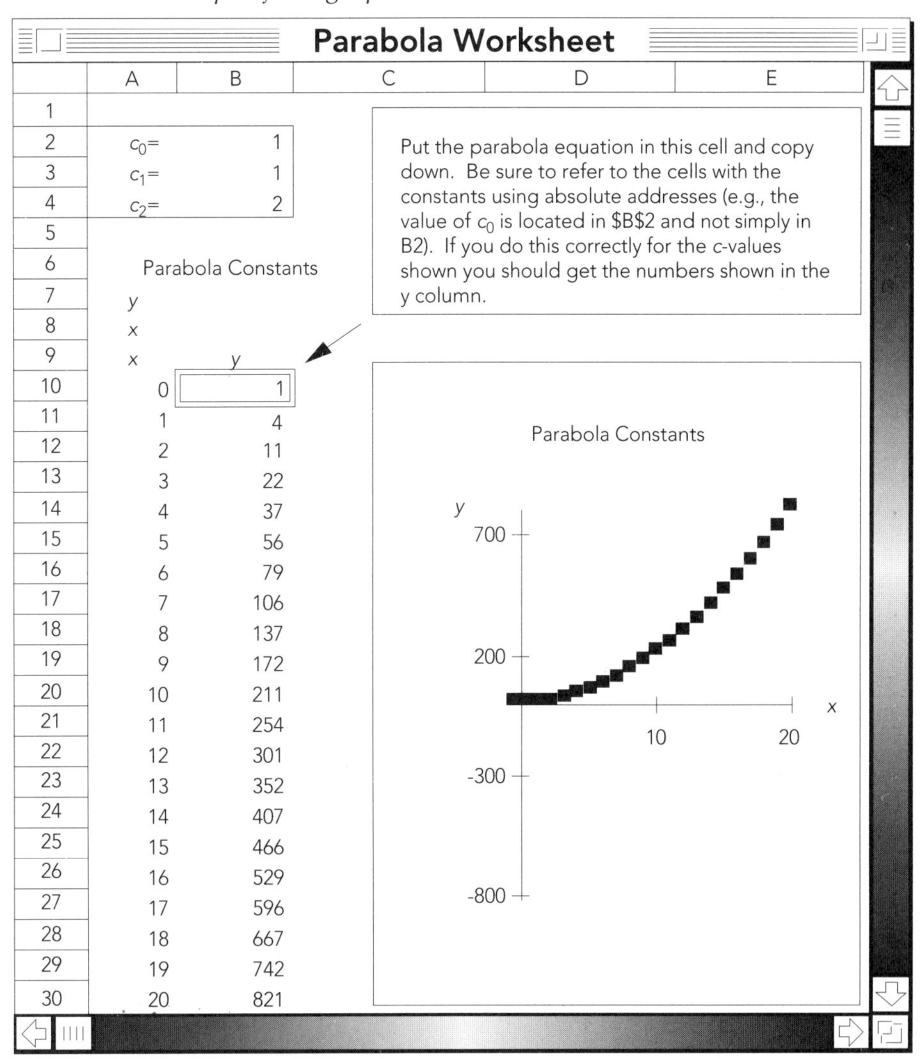

Fig. 4.13. Copy of the Parabola Worksheet screen.

4.6.1. Activity: Parabolas for Different C_0, C_1, and C_2 Values

a. First let's explore the effect of the value of c_0 on the location of the parabola. Examine the equation for the parabola and predict what you think the effect of changing the value of c_0 will be. Suppose it is changed to +100? To –100? What will happen in each case? Explain the reasons for your prediction.

b. Test your prediction by changing the values of c_0 from +1 to +100 and then to –100. Sketch graphs for these three different values of c_0 in the space below using solid dotted and dashed lines and marking the appropriate values of c_0 on each sketch.

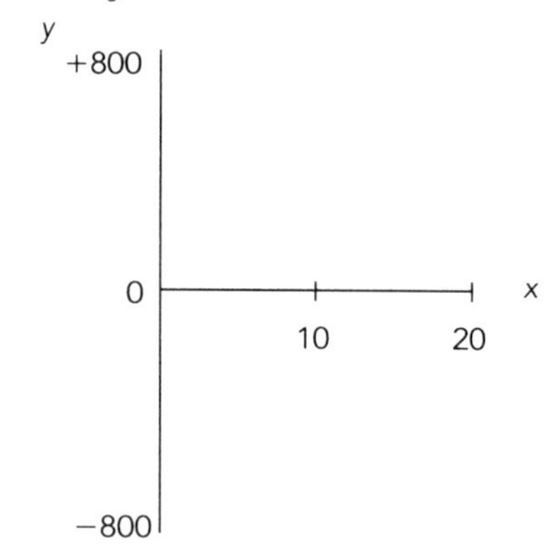

c. Describe in words what c_0 has to do with the y–intercept of the graph. Does the basic shape (curvature) of the parabola change in any way? **Note:** If x represented time and y represented position, then the y-intercept would be the position of an object at time $t = 0$.

d. Find the derivative $\frac{dy}{dx}$ of the parabola equation

$$y = c_2x^2 + c_1x + c_0$$

e. Show that this derivative is simply the constant c_1 at the y-intercept (i.e., when x=0).

f. Does c_1 actually determine the slope of the parabola when $x = 0$? To check this, we suggest that you return the values of the constants to +1, +1, and +2 and then try changing c_1 from +1 to +30 and then to

–30. Sketch graphs for these three different values of c_0 in the space below using solid dotted and dashed lines and marking the appropriate values of c_1 on each sketch.

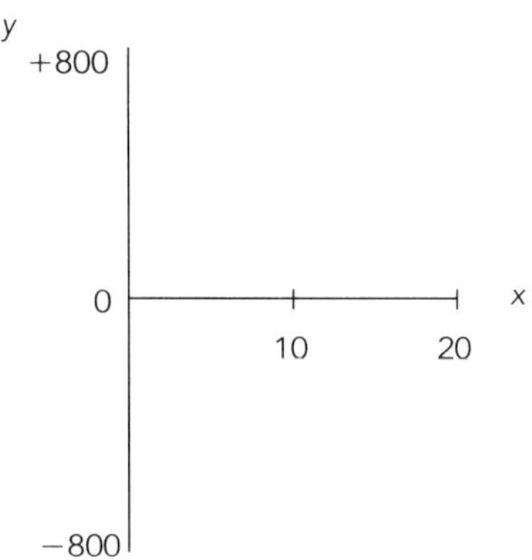

g. Describe in words what c_1 has to do with the slope of the graph at the y-intercept. Does the basic shape (curvature) of the parabola change in any way?

> **Note:** If x is time and y is position, then the derivative of the x vs. t equation represents the velocity of the object as a function of time.
>
> $$\frac{dy}{dx} \Rightarrow \frac{dx}{dt} = v$$
>
> The slope at the y-intercept would be the initial velocity of an object.

h. Explore the effect of the value of c_2 on the shape of the parabola. To do this, it is suggested that you return the values of the constants to +1, +1, and +2 and then try changing c_2 from +2 to –2 and then to +4. Sketch graphs for these three different values of c_2 in the space below using solid dotted and dashed lines and marking the appropriate values of c_2 on each sketch.

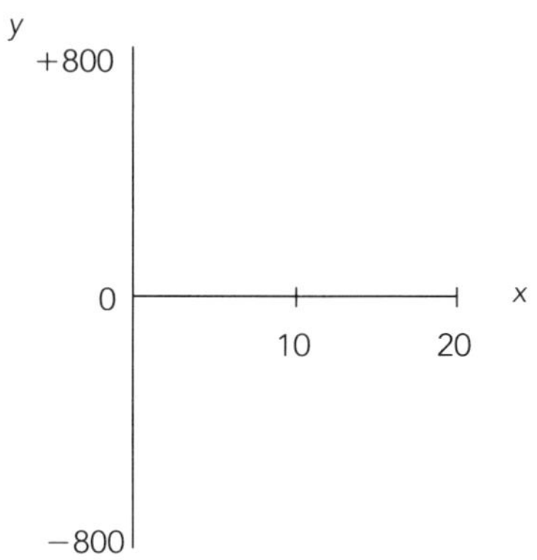

i. Describe in words what c_2 has to do with the *shape* of the graph. Does the basic shape (curvature) of the parabola change in any way? **Note:** If x represented time and y represented position, then the shape of the graph with a higher curvature would represent higher magnitudes of acceleration. The upside down parabola would represent a motion with negative acceleration.

Note: Save your Parabola Worksheet since it might help you complete the next activity more rapidly.

4.7. MODELING: FITTING AN OBSERVED MOTION WITH AN EQUATION

Can you find the appropriate values of the constants c_2, c_1, and c_0 in the parabola equation

$$y = c_2x^2 + cx + c_0$$

so that its shape and location are just right to fit a real position vs. time graph? You can use the data that you collected in Activity 3.7.1. for an accelerated cart speeding up as recorded by a video camera.

To complete this next activity you will need:

- 1 microcomputer
- 1 spreadsheet software
- 1 modeling worksheet (optional)

Recommended group size:	2	Interactive demo OK?:	N

If you don't have a modeling worksheet available, you can create your own. The modeling worksheet is similar to the parabola worksheet used in the last activity. However, it has an additional column for experimental data and the graph is an overlay of the experimental data points with a line representing the calculations based on an equation for a parabola. Once the y and x labels are replaced with position in meters, x (m) and time in seconds, t (s), and the times for the cart measurements are entered, the spreadsheet should look something like that shown in the following illustration.

With x and t taking the roles of y and x respectively the parabola equation becomes

$$x = c_2t^2 + c_1t + c_0$$

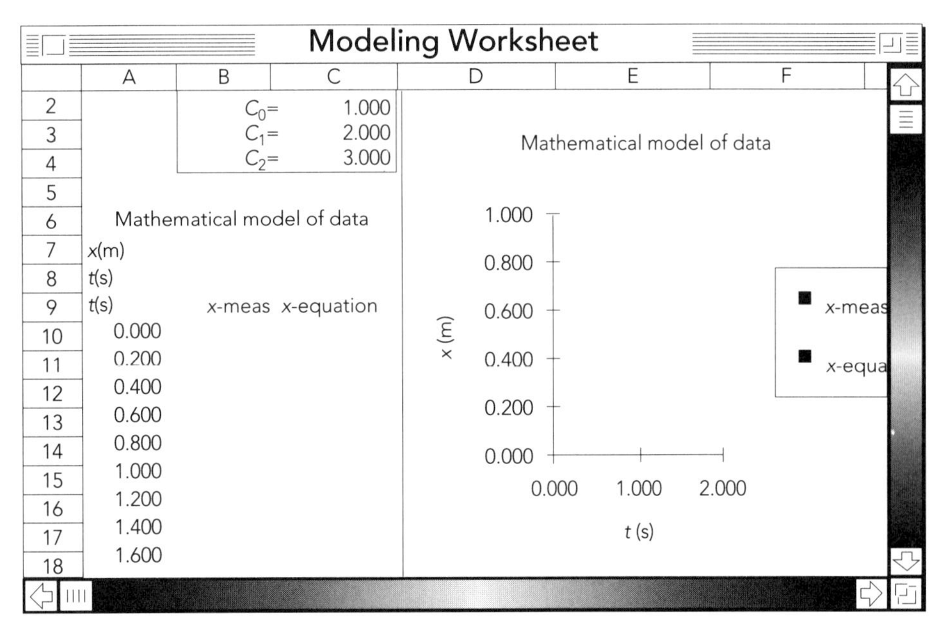

Fig. 4.14. A partially configured modeling worksheet.

4.7.1. Activity: Modeling the Cart Position Data

a. Examine your data in Unit 3, Activity 3.8.1c. What is the position of the cart in meters at $t = 0.00$ seconds? What is its average velocity in the first time interval—that is, from $t = 0.00\,\text{s}$ to $t = 0.200\,\text{s}$?

b. Open up a spreadsheet or a modeling worksheet and

1. Enter the indicated title, labels, and values of time in the first column of the spreadsheet.
2. Enter the measured values for position into the x-measured column in the worksheet.
3. Enter some arbitrary guesses for the values of the constants in the first three rows of the spreadsheet.
4. Enter the equation for the parabola in the x-equation column of the spreadsheet. (As always in modeling, be sure to call on the values of the constants using absolute references.)
5. Select the values to be graphed as an overlay graph of x-meas vs. t and x-equation vs. t. Make the x-equation data into a line on the graph (instead of data points).
6. Finally, adjust the values of the constants until the curve passes very near or through most of the experimental data points. Affix

a printout of your matched graphs and the values of the constants in the space that follows.

PRINTOUT OF GRAPH GOES HERE

c. Substitute the constants into the parabolic equation for x as a function of time to write down the equation that describes the motion you studied.

d. How does the value of c_0 compare with the initial position of the cart?

e. How does the value of c_1 compare with the initial velocity of the cart?

f. As you will see in the next activity, the value of c_2 should be about half of the cart acceleration. Is it?

THE KINEMATIC EQUATIONS

4.8. 1D KINEMATIC EQUATIONS FOR CONSTANT ACCELERATION

So far, you should have concluded that some of the motions have had accelerations that are more or less constant. *There is a standard set of equations (which can be derived using the principles of calculus) that describe the motion of an object that undergoes constant acceleration.* These equations are called the *kinematic equations* and they are derived in most standard physics textbooks. By re-examining the graphs you have sketched that describe objects moving with constant acceleration, and by using the definitions of instantaneous velocity and acceleration, you can verify that the kinematic equations describe uniformly accelerated motion. We will use the following symbols for this exercise:

x = position along the x-axis (which can vary with time)

v = instantaneous velocity along the x-axis (which can also vary with time)

a = constant acceleration along the x-axis (it does not vary in time because we have chosen to consider only those motions for which a is constant.)

x_0 = initial position at $t = 0$

v_0 = initial velocity component along the x-axis at $t = 0$

t = the time elapsed time since the object was at x_0.

Beware: The kinematic equations you are about to derive in the space below *only apply when an object undergoes constant acceleration.*

There are four constant acceleration kinematic equations commonly found in physics textbooks. The most fundamental kinematic equation is the equation describing x as a function of t when the initial position is x_0 and the initial velocity is v_0.

Kinematic Equation #1:

$$x = \frac{1}{2}at^2 + v_0 t + x_0 \qquad (4.2)$$

This equation indicates that a graph showing the position as a function of time of any motion with constant acceleration is a parabola of some sort. In fact, you should have verified the experimental validity of this equation in the mathematical modeling exercise you did in Activity 4.7.1. In that activity you should have determined that a graph of a parabola like that shown in Equation 4.2 and a graph of your data are almost identical. **Note:** All other kinematic equations can be obtained from the fundamental equation and the definitions of instantaneous velocity and acceleration.

This equation can be derived mathematically by using the formal definitions of instantaneous velocity and acceleration. Instead of deriving the kinematic equation formally, we want you to use Kinematic Equation #1 to verify that it represents a motion with constant acceleration. Although this verification is not entirely rigorous mathematically, your observations coupled with the verification exercise which follows should convince you that this first kinematic equation is reasonable *both* experimentally and theoretically.

4.8.1. Activity: Verifying the Kinematic Equations Mathematically

a. Use the definition of instantaneous velocity ($v \equiv dx/dt$) and take the derivative of x with respect to time to show that if kinematic equation #1 is valid, then the second kinematic equation would be

Kinematic Equation #2:

$$v = at + v_0 \tag{4.3}$$

b. Use the definition of instantaneous acceleration ($a \equiv dv/dt$) to show that the second kinematic equation is valid (i.e., that a = constant) by taking the derivative of equation #2 with respect to time. **Note:** Since this second equation is valid, we can argue that the first one it came from is also valid.

The other two kinematic equations given in the text can be derived by combining Equations 4.2 and 4.3. In fact, we will do this in section 4.9. However, since all constant acceleration problems can be solved using Equation 4.2 for $x(t)$ and the rules for taking a derivative to get Equation 4.3, we do not recommend trying to memorize the other equations listed in textbooks.

4.9. USING THE KINEMATIC EQUATIONS

The kinematic equations can be very useful in helping to do calculations that enable you to predict with fair accuracy the positions and velocities of constantly accelerated objects. Also, if you know enough about how an object is moving, you can calculate its acceleration. In this section you are going to practice using some elements of formal problem solving.† These elements are:

Part 1: Motion Diagrams and Graphs
Part 2: Tables and Equations
Part 3: Algebra and Substitution
Part 4: Computation and Unit Checks

In order to learn about these elements of formal problem solution we

† This approach was developed by Robert Morse, a teacher at St. Albans School in Washington, DC.

would like you to consider two different types of kinematic problems. For each type of problem, an example will be presented and worked, and then you will be asked to work a practice problem.

TYPE ONE PROBLEMS

In type one problems we usually know the time an object is accelerating and some combination of other variables. In the case of type one problems Kinematic Equation #1 is the key equation.

Type One Kinematics Problems

Example Problem About a Runner:

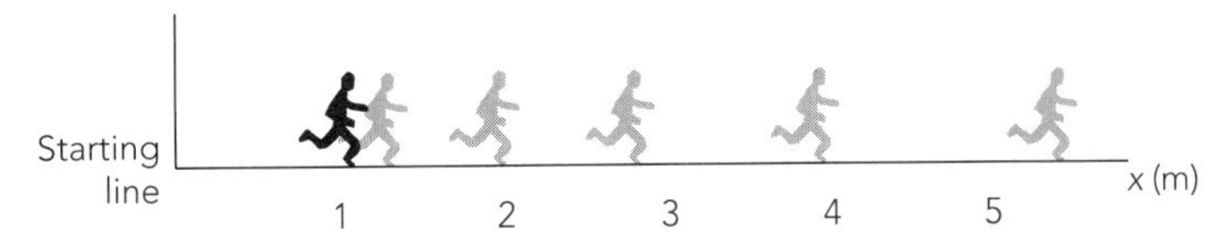

Fig. 4.15.

Carl Lewis is hoping for a record finishing time in the 100 meter run. During the first meter Lewis gets off the blocks into his upright running position and is already moving at a speed of 3.4 mi/h. During the next four meters of his run he accelerates at a constant rate to his final running speed. Yolanda, a young sport scientist who is studying his acceleration clocks the time that it took Lewis to run from 1.0 m to 5.0 m in 0.69 s. What is his acceleration?

Practice Problem About a Jet Boat: The Rogue River in Oregon is so rocky in spots that jet boats have become a popular way for tourists to see wild stretches of the river. A system of air jets thrusting downward keep the boat suspended above the water and a second system of air jets move it horizontally. A jet boat driver comes around a bend in the river moving due west at a speed of 12 m/s when she discovers that a large Douglas fir tree has fallen across the river 24 meters in front of her. She quickly reverses her horizontal jets so that they deliver a constant acceleration in an easterly direction. The jet boat slows down considerably and reaches the log after 40 seconds. What is its acceleration? Does it stop in time?

On the pages that follow all four parts of a suggested solution to the runner problem are described. After each element is explained you will have a chance to practice using these elements to solve the jet boat problem in which a jet boat driver avoids a collision with a fir tree.

Part 1: Motion Diagrams and Graphs

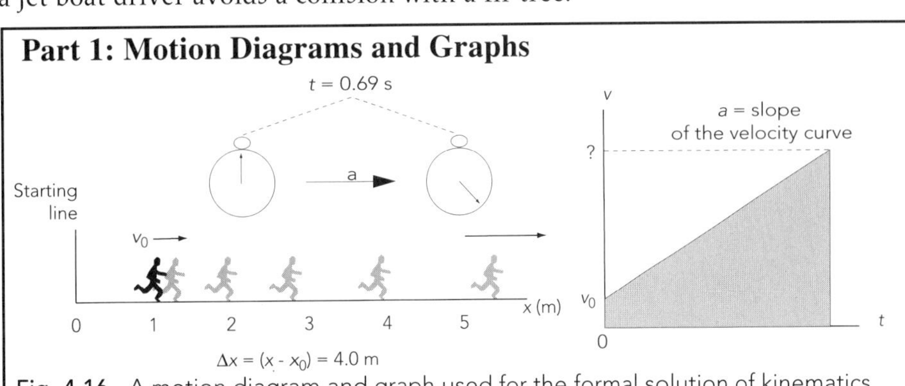

Fig. 4.16. A motion diagram and graph used for the formal solution of kinematics problems.

Notice that we depict the runner at his known positions at the beginning and end of the time interval. The arrows associated with his velocity are shown at those two locations and have different lengths–longer for greater velocity. The direction of the acceleration is shown with an arrow also. All the relevant variables are indicated by their symbols. The diagram is an abstract representation of the problem. A second representation is the sketched graph. The slope represents the acceleration, the initial velocity and the time interval are also shown. It is sometimes possible to solve the problem merely by constructing a graph.

4.9.1. Activity: Jet Boat Problem Motion Diagram and Graph

a. Draw a motion diagram to represent practice kinematics problem about the jet boat.

b. Sketch a velocity vs. time graph for the jet boat assuming it has a constant acceleration.

Part 2: Table of Values and Equations

List all the values given in the problem and convert to a common set of units you plan to use. This is usually based on the International System of Units which is abbreviated SI (based on the French version of the name). This system is sometimes referred to as the MKS system to reflect its use of meters, kilograms and seconds as fundamental units.

In this example the runner's initial and final position was given and the time required to move between those positions was given. Thus, you will need to use Kinematic Equation #1.

$$x = \frac{1}{2}at^2 + v_0 t + x_0$$

The table of values, conversions, and the derivation of this equation are shown as follows.

Part 2: Table and Unit Conversions	**Part 2: Runner Equations**
$x_0 = 1.0$ m $x = 5.0$ m $v_0 = 3.4$ mi/h $= (.447)(3.4)$ m/s $= 1.52$ m/s (1 mi/h $= 0.447$ m/s) $t = 0.69$ s	Only one equation is needed for this situation but it must be solved for acceleration $x = \frac{1}{2}at^2 + v_0 t + x_0$

4.9.2. Activity: A Table and Equations for the Jet Boat

a. Make up a table of values to represent practice problem about the jet boat. Be sure to show any unit conversions, if needed.

b. Write down or derive the basic equations needed to find the acceleration of the jet boat as a function of the distance it moves and its initial position and velocity and the time the motion takes. Do not solve the equation(s) for acceleration quite yet.

Part 3: Runner Algebra and Substitution

The next step is to solve the equation algebraically for the quantity of interest. In this problem it is the acceleration, a, but in other problems the acceleration may be known and a velocity or the distance moved might be of more interest.

Runner Algebra and Substitution

Solve for a:

Since $x = \frac{1}{2}at^2 + v_0 t + x_0$, then $a = \frac{2[(x - x_0) - v_0 t]}{t^2}$.

$$\text{so } a = \frac{2[(5.0 - 1.0) - (1.5)(0.69)]}{0.69^2} = 13\,\text{m/s}^2$$

ANSWER (with proper significant figures): $a = 13\,\text{m/s}^2$

4.9.3. Activity: Jet Boat Algebra and Substitution

a. Solve the equation obtained in Part 2 for the acceleration of the practice problem jet boat algebraically.

b. Substitute the values in appropriate units into the equation and calculate the acceleration.

Part 4: Units Check

The final step is to see that the units represented on the right hand side of your final equation used in Part 3 and the units on the left hand side are the *same.* For example, for the case of our stopping cubicle:

Runner Units Check

$$[\text{m/s/s}] = \frac{[\text{m}] - [\text{m/s}][\text{s}]}{[\text{s}^2]} = \frac{[\text{m}]}{[\text{s}^2]}$$

4.9.4. Activity: Jet Boat Units Check

Check to see that the units on both sides of the final equation you are using to find the acceleration of the jet boat are the same.

TYPE TWO PROBLEMS

In type two problems we usually know both the initial and the final velocity of an object and either the acceleration of the object or the distance over which it moves. In the case of type two problems, you can either use a combination of Kinematic Equation #1 and the definition of acceleration or you can use Kinematic Equation #3 which will be derived later in this section.

Type 2 Kinematics Problems

Example Problem 2:

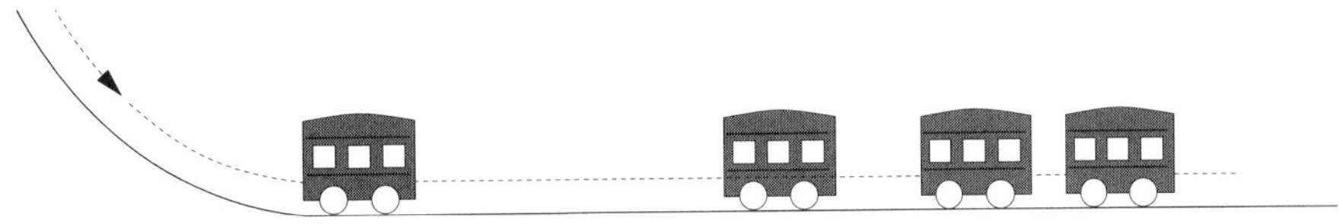

Fig. 4.17.

A cage holding four people at the Cedar Point Amusement Park has been accelerated as the result of a vertical free fall. It changed direction on a track and is coasting horizontally with an initial velocity of 40 m/s when the brakes are applied. The cart slows down at a constant acceleration and stops in a distance of 18 m. What is its acceleration due to the braking action?

Practice Problem 2: A Boeing 757 jumbo jet with 400 passengers and full fuel tanks has a constant acceleration at full throttle of 2.0 m/s/s and must be going 170 mi/h to lift off. You have to decide as a Federal Aviation Administration safety officer whether to allow United Airlines to operate the Boeing 757 on a new sea level airfield with a runway length of 1.00 miles. Will you allow United to operate B757s on the new airfield?

On the pages that follow all four parts of a solution to the example problem on the stopping cage are described. After each element is explained you will have a chance to practice using it to solve the practice problem in which a Boeing 747 jumbo jet takes off from an airfield that might be too short.

Part 1: Motion Diagrams and Graphs

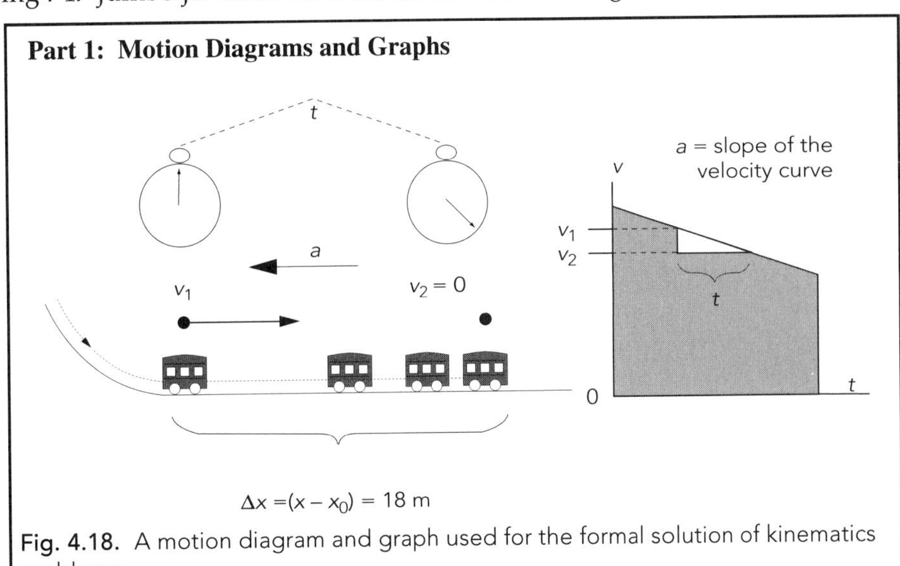

Fig. 4.18. A motion diagram and graph used for the formal solution of kinematics problems.

Notice that in Figure 4.18 we see the cart twice—once at the beginning of the time interval, in the state of motion described by v_1, and once at the end of the time interval in the state of motion described by v_2. The arrows associated with each object have different lengths—longer for greater velocity. All the relevant variables are indicated by their symbols. The diagram is, in effect, an abstract representation of the problem. A second representation is the sketch graph on the right. The slope represents the acceleration, the area represents the displacement, and both velocities and the time interval are also shown. It is sometimes possible to solve the problem merely by constructing a graph.

4.9.5. Activity: A Motion Diagram and Graph for the Jet Plane

a. Draw a motion diagram to represent practice problem 2 about the Boeing 757.

b. Sketch a velocity vs. time graph for the Boeing assuming it is taking off with a constant acceleration.

Part 2: Table of Values and Equations

List all the values given in the problem and convert to a common set of units you plan to use. This is usually based on the SI system in which lengths are in meters and time in seconds.

In this example the initial and final velocity of the cage was given and the distance moved between them, $(x - x_0)$ was given instead of the time. Thus, you will need to use Kinematic Equation #1 and its first derivative to eliminate time. Doing this gives a very important additional kinematic equation relating velocities, distance moved, and acceleration. This equation is the third kinematic equation.

Kinematic Equation #3:

$$v^2 = v_0{}^2 + 2a(x - x_0) \tag{4.4}$$

The table of values, conversions, and the derivation of this third kinematic equation are shown as follows.

Part 2: Table and Unit Conversions

$v_i = v_0 = 40$ m/s

$v_f = v = 0.0$ m/s

$(x - x_0) = 18$ m

$a = ?$

Part 2: Equations for the Stopping Cage

$$x = \frac{1}{2}at^2 + v_0t + x_0 \text{ and } v = \frac{dx}{dt} = at + v_0,$$

$$\text{so} \quad t = \frac{v - v_0}{a}, \quad \text{or}$$

$$x - x_o = \frac{1}{2}a\left(\frac{v - v_0}{a}\right)^2 + v_0\left(\frac{v - v_0}{a}\right)$$

$$= \frac{1}{2a}(v^2 - 2vv_0 + v_0^2) + \frac{1}{a}(vv_0 - v_0^2)$$

$$x - x_0 = \frac{1}{2a}(v^2 - v_0^2)$$

so finally we get the third kinematic equation:

$$v_0^2 = v_0^2 + 2a\,(x - x_0)$$

4.9.6. Activity: A Table and Equations for the Jet Plane

a. Make up a table of values to represent the practice problem about the Boeing 757. Be sure to show any unit conversions.

b. Write down or derive the basic equations needed to find the horizontal distance the Boeing moves as a function of its acceleration and its initial and final velocities.

Part 3: Algebra and Substitution

The next step is to solve the equation algebraically, if needed, for the quantity of interest. In this problem it is the acceleration, a, but in other problems the acceleration may be known and a velocity or the distance moved might be of more interest.

Stopping Cage Algebra and Substitution

Solving Kinematic Equation for the Stopping Cage for acceleration gives

$$a = \frac{(v^2 - v_0^2)}{2(x - x_0)} = \frac{(0^2 - 40^2)\mathrm{m}^2/\mathrm{s}^2}{2(18)\mathrm{m}}$$

ANSWER (with proper significant figures) $a = -44\ \mathrm{m/s^2}$

4.9.7. Activity: Algebra and Substitution for the Jet Plane

a. Solve the equation obtained in Part 3 for the horizontal distance of the Boeing as it moves before it reaches take off velocity.

b. Substitute the values in appropriate units into the equation and calculate the distance moved by the B757.

c. As an FAA safety officer would you allow United Airlines to operative B757s on the new airfield? Explain.

Part 4: Units Check

The final step is to see that the units represented on the right hand side of your final equation used in Part 3 and the units on the left hand side are the same. For example, for the case of our stopping cage:

Units Check for Stopping Cage

$$[\mathrm{m/s/s}] = \frac{[(\mathrm{m/s})^2]}{[\mathrm{m}]} = \frac{[\mathrm{m}^2/\mathrm{s}^2]}{[\mathrm{m}]} = \frac{[\mathrm{m}]}{[\mathrm{s}^2]} = [\mathrm{m/s/s}]$$

4.9.8. Activity: Units Check for the Jet Plane

Check to see that the units on both sides of the equation agree.

Name ______________________ Section __________ Date __________

UNIT 5: ONE-DIMENSIONAL FORCES, MASS, AND MOTION

The Apollo/Saturn V space vehicle carrying Apollo 11 astronauts Neil A. Armstrong, Michael Collins, and Edwin E. Aldrin, Jr., lifted off at 9:32 A.M. EDT on July 16, 1969. This was our nation's first manned lunar landing mission. The 36-foot-high vehicle generated a thrust of seven and one-half million pounds during liftoff. It lumbered off the launch pad very slowly at first, and then picked up speed rapidly. Its velocity may have increased at slightly more than a constant rate during the early stages of take off. This increasing acceleration was probably followed by a decreasing acceleration even if the ejected fuel created a constant thrust force on the rocket. Yet, the rocket's motion and escape from the Earth's gravitational attraction happened in accordance with Newton's Laws of motion. How is it possible for the rocket to give itself a constant thrust force and have an acceleration that is not constant? As you study the fundamental relationships between one-dimensional net *force, mass, and motion in this unit, you should be able to answer this question.*

UNIT 5: ONE-DIMENSIONAL FORCES, MASS, AND MOTION

If you find the study of motion difficult, reflect that it took mankind . . . over sixteen hundred years to reach a clear understanding of motion; you should hardly be impatient if it takes you several weeks.
Eric Rogers (1961)

No one must think that Newton's great creation [the three laws of motion] can be overthrown. . . . His clear and wide ideas will forever retain their significance as the foundation on which our modern conceptions of physics have been built. Albert Einstein (1948)

OBJECTIVES

1. To devise a method for applying a constant force to an object.
2. To find a mathematical relationship between force and motion.
3. To devise a force scale to measure one, two, three, etc. units of force.
4. To understand how different forces can act to make up a combined force.
5. To develop a definition of mass in terms of an object's motion under the influence of a force.
6. To combine all of the observations and develop statements of *Newton's First and Second Laws of Motion* for one-dimensional motion with negligible friction.

5.1 OVERVIEW

So far in your study of one dimensional motion you have learned to observe and describe motion in several ways. The next step is to study the causes of motion.

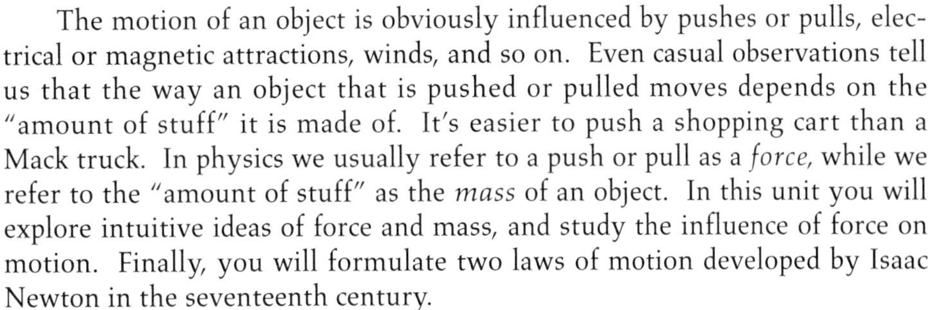

The motion of an object is obviously influenced by pushes or pulls, electrical or magnetic attractions, winds, and so on. Even casual observations tell us that the way an object that is pushed or pulled moves depends on the "amount of stuff" it is made of. It's easier to push a shopping cart than a Mack truck. In physics we usually refer to a push or pull as a *force,* while we refer to the "amount of stuff" as the *mass* of an object. In this unit you will explore intuitive ideas of force and mass, and study the influence of force on motion. Finally, you will formulate two laws of motion developed by Isaac Newton in the seventeenth century.

Fig. 5.1.

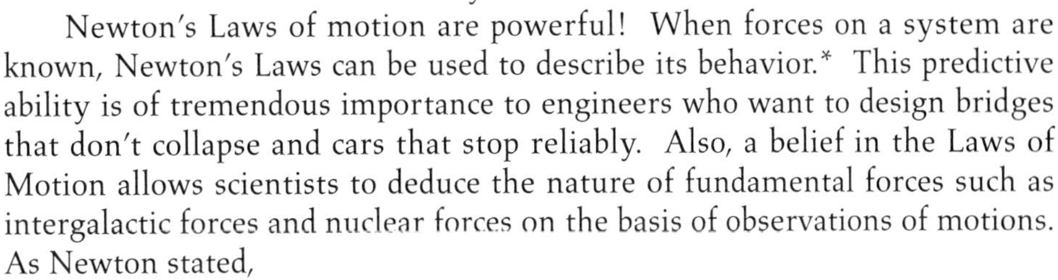

Newton's Laws of motion are powerful! When forces on a system are known, Newton's Laws can be used to describe its behavior.* This predictive ability is of tremendous importance to engineers who want to design bridges that don't collapse and cars that stop reliably. Also, a belief in the Laws of Motion allows scientists to deduce the nature of fundamental forces such as intergalactic forces and nuclear forces on the basis of observations of motions. As Newton stated,

> . . . the phenomena of motions [can be used] to investigate the forces of Nature, and then . . . these forces [can be used] to demonstrate other phenomena...the motions of the planets, the comets, the moon and the sea.

Note: The classical laws of motion that we will develop in this unit provide for all practical purposes "exact" descriptions of the motions of everyday objects traveling at ordinary speeds. During the early part of the twentieth century two new theories were developed—quantum theory, which describes motion in the atomic realm, and relativity, which describes objects moving extremely fast. Once you master the classical description of ordinary motions, it is exciting indeed to see how these laws are modified so that they will also describe very small objects or objects moving at speeds close to that of light.

---

* Often when forces are known, the actual position and velocity of the system can be predicted within the limits of experimental uncertainty. However, there are some systems that exhibit chaotic behavior that can be described but not predicted. In Unit 15 you will study a chaotic system.

FORCE AND MOTION

5.2 MOTION FROM A CONSTANT FORCE

In your previous study of motion in this course you concentrated on describing motion rather than on understanding its causes. From your experiences, you know that force and motion are related in some way. For example, to start a bicycle moving, you have to push down on the pedals, or to start a wagon moving you have to pull on it. What kinds of forces lead to steady motion? To changes in motion?

Before you can study the relationship between force and motion, you must be able to devise a useful definition of force and develop reliable ways to measure it. Then you can investigate these relationships by applying forces of different strengths to various objects. You can begin by figuring out how to apply a constant force to a person who can slide along a smooth floor or roll along on a low friction cart. Then you can measure the motion of the person under the influence of a constant force using a computer-based laboratory system.

In order to do the activities in this section you will need some but not necessarily all of the following items to investigate the creation of a constant force:

- 1 rod
- 1 table clamp
- 1 ruler
- 1 meter stick
- 1 large rubber band, #117 (3.5″ x 0.75″)
- 1 large spring scale, about 15 kg
- 1 mass pan, 1 kg
- 2 masses, 1 kg
- 3 masses, 2 kg

These items can be used to relate force and motion:

- 1 computer-based laboratory system
- 1 ultrasonic motion sensor
- 1 motion software
- 1 large plastic garbage bag
- 1 Kinesthetics cart

Recommended group size:	2	Interactive demo OK?:	Y

What Is Force?

What is force and how is it measured? The word force is a very common part of everyday language. One of the major tasks in this unit is to help you move in stages from an informal understanding of the meaning of the term force as a push or pull to a more precise, quantitative definition that is useful in relating force to motion.

5.2.1. Activity: Ideas about Force

Attempt to define the word force in your own words. What are some examples of forces? How might you measure how large a given force is?

Creating a Constant Force

In order to explore the relationship between force and motion, you should try to figure out how to apply a constant force to an object. You can use any of the equipment listed for this section or any other common items you have available.

5.2.2. Activity: Applying a Constant Force

Devise a method for pulling on an object with a constant force. Explain your method and also explain why you believe that the force is constant.

Predicting Motion from a Constant Pull

Suppose you exert a fairly large constant pull on a person? We are interested in having you track the motion of a person sitting on top of a large garbage bag as well of the motion of a person riding on a low friction cart. A computer based laboratory system can be set up to track the motion with a motion sensor placed behind the person being pulled. These situations are shown in Figure 5.2.

Fig. 5.2. Being pulled with a constant force under two circumstances–(a) sliding along a smooth floor on a plastic garbage bag and (b) rolling along a level floor on a low-friction Kinesthetic cart.

5.2.3. Activity: Predicting the Velocity of a Person Being Pulled with a Constant Force

a. Consider the situation in Figure 5.2a. What do you predict you might see for a velocity vs. time graph if the person starts from rest and slides along the floor while being pulled away from a motion sensor with a constant applied force? Sketch the shape of the predicted graph in the space below.

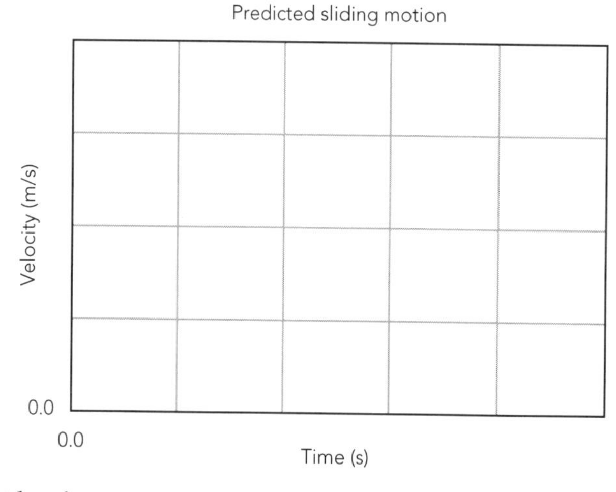

b. Consider the situation in Figure 5.2b. What do you predict you might see for a velocity vs. time graph if the person starts from rest and rolls along the floor while being pulled away from a motion sensor with a constant force? Sketch the shape of the predicted graph in the space below.

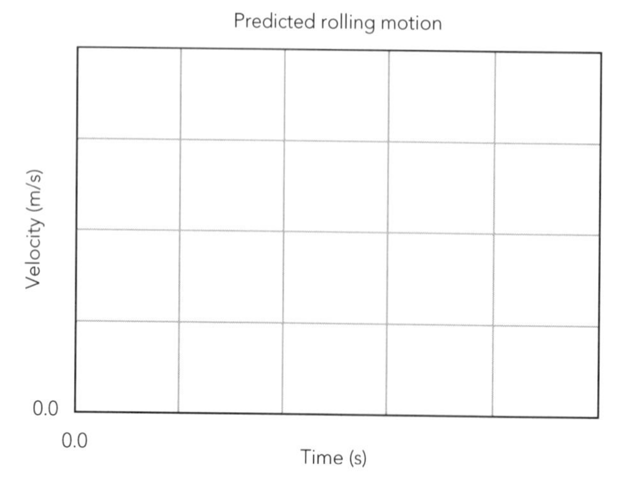

c. Explain the reasons for your predictions. If you predict that the two graphs will have the same shape, explain why. If you predict different shapes, explain why you expect the shapes to be different.

Observing Motion from a Constant Pull

You can work with the rest of the class or with one or more partners to create and record the sliding and rolling motions with a constant force using a computer-based laboratory system and motion detector. We suggest you use a large rubber band or spring scale stretched out to a constant distance to create a constant pulling force. If the person being pulled holds a meter stick, the puller can try to keep the stretch fairly constant. The person should be pulled away from the motion sensor. These motions take some practice to create. To get reliable measurements, apply enough constant force in each case to get velocities of 0.5 m/s or more for a time period of about 5 seconds.

5.2.4. Activity: Observing the Velocity of a Person Being Pulled with a Constant Force

a. Create the sliding situation depicted in Figure 5.2a and observe what the velocity is as a function of time. Fill in the horizontal and vertical axis values on the following graph frame and sketch the observed graph.

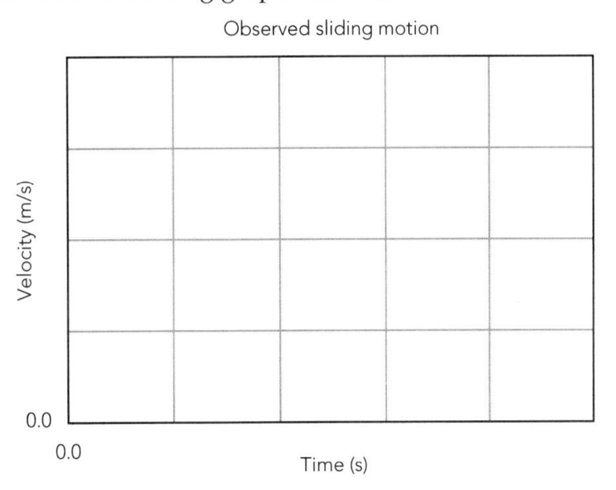

b. Examine the overall shape of the graph for the sliding motion. (Ignore the smaller bumps associated with the wobbling spring or rubber band.) Is the velocity zero, constant, or changing? Is the acceleration zero, constant, or changing? Explain what characteristic of the graph shape supports your description.

c. Explain how you created the constant force. Describe the pulling method used. If you stretched rubber band or spring, what was the amount of the stretch?

d. Create the rolling situation depicted in Figure 5.2b and observe what the velocity is as a function of time. Fill in the horizontal and vertical axis values on the following graph frame and sketch the observed graph.

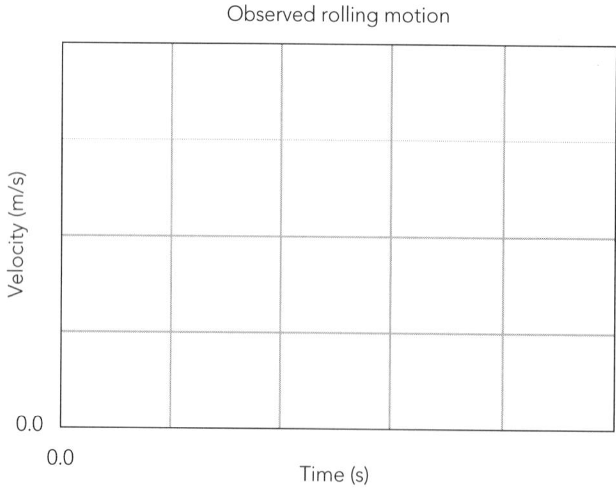

e. Examine the overall shape of the graph for the rolling motion. (Ignore the smaller bumps associated with the wobbling spring or rubber band.) Is the velocity zero, constant, or changing? Is the acceleration zero, constant, or changing? Explain what characteristic of the graph shape supports your description.

f. How did the observations you reported in a. and d. compare with your predictions?

g. State a general rule based on your observations for the relationship between a constant applied force and velocity when *friction is significant,* for example, when an object slides.

h. State a general rule based on your observations for the relationship between a constant applied force and acceleration when the friction is low.

i. Based on your observed velocity vs. time graph for the low friction rolling motion resulting from a steady pull, sketch an acceleration vs. time graph for the time period during which the pulling force was roughly constant.

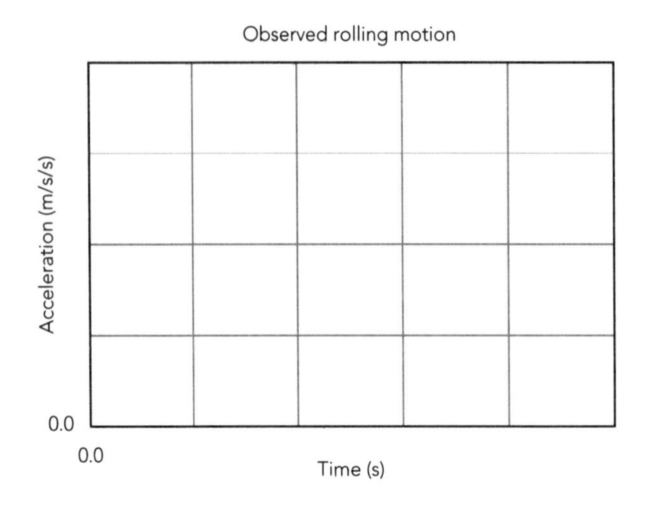

Rolling vs. Sliding Motions

As you probably observed, a constant force leads to a constant velocity when an object is sliding. On the other hand, an object such as a low-friction cart with a person riding on it moves at a constant acceleration under the influence of a constant force. This is a surprising observation to most people! The sliding motion involves a significant amount of friction while the rolling does not. For the rest of this unit, you will be working with rolling objects that have only a small amount of friction to discover Laws of Motion in the absence of friction. In Unit 6 you will return to the question of how friction can be incorporated into the laws of motion.

5.3 COMBINING EQUIVALENT FORCES

In the last section you should have discovered that a single constant force applied to a object that rolls without much friction causes it to move with a constant acceleration. What happens to the acceleration of a rolling object when the force doubles or triples? What if the force is not constant? What happens to the object's acceleration then?

Before you can proceed with further investigations of the relationship between force and motion, you need to explore reliable ways to measure arbitrary forces. *Being able to produce and combine equivalent forces will enable you to define a force scale and understand how forces of arbitrary strengths can be measured and created.*

We'd like you to work in small groups to investigate how to find equivalent forces and combine them. You will use small low-friction dynamics carts. Thus, you need to learn to work with a smaller forces than the one used to pull a student riding on a Kinesthetics cart.

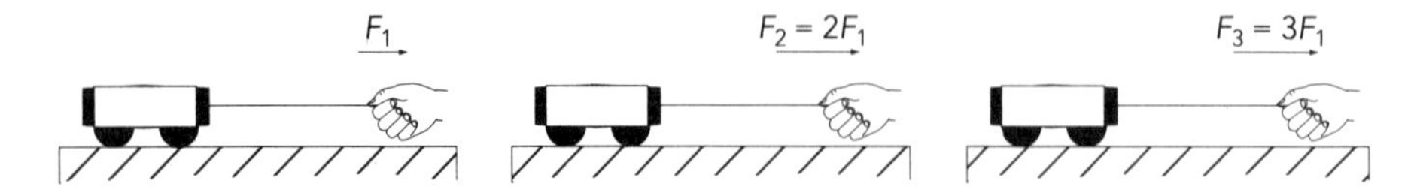

Fig. 5.3. Given a force F_1, how can we apply double or triple that force?

The activities in this section on combining equivalent forces are not completely specified. Although the following items should be available, you might not need to use all of them:

- 2 table clamps
- 2 rods
- 1 ruler
- 6 rubber bands, #14
- 2 mass pans, 50 g
- 2 masses, 50 g
- 5 masses, 100 g
- 1 spring scale, 10 N
- 2 lengths of lightweight string, 24″
- 1 small low-friction dynamics cart
- 2 masses, 500 g
- 1 smooth ramp or level surface 1–3 meters long

Recommended group size:	2	Interactive demo OK?:	N

Equivalent Forces

One strategy for creating double and triple forces is to create and combine equivalent forces. You should start this investigation by defining a force unit by pulling a single rubber band out to some predetermined length that you choose. You could name this amount of force after yourself or make up another name for it.

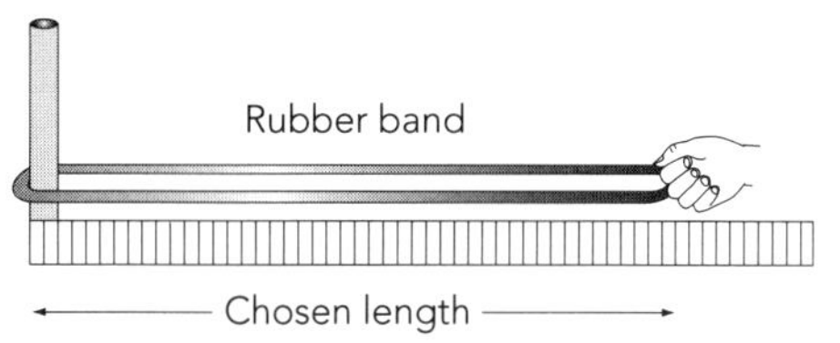

Fig. 5.4. Pulling a rubber band out to a chosen length.

5.3.1. Activity: Creating Equivalent Forces

a. Pick out one of the #14 rubber bands as your standard rubber band. You may want to identify it by marking it with a pen or pencil. Now use your rubber band to define your own unit of constant force. Explain how your unit of force is created. In other words, if you were to give your rubber band to someone else, explain how they could use it to pull on something with your unit of constant force.

b. Consult with your partner(s) and decide what you want to call your unit.

c. Suppose we define an equivalent force as a force that accelerates a cart in exactly the same way that your unit of force does. Consult with your partner(s) and think of as many different techniques as possible to create an *equivalent force* using different objects than your special rubber band. Describe three or more of these techniques.

d. Now consult with your partners and think of two or more ways to test whether a proposed equivalent force is actually equivalent to your force unit.

e. Now create what you think is an equivalent force. Does the "equivalent" force pass the tests you have devised in part d. for equivalency? Explain.

Evaluating Alternatives for Testing Equivalent Forces

Sometimes when there are alternative ways to accomplish a goal some seem better than others. You should discuss your ideas and findings with your classmates and then choose what you think is the best technique for creating a force that is equivalent to your unit in terms of its ability to accelerate a cart in the same way that your force did.

5.3.2. Activity: Equivalent Force Techniques

a. Are there any other tests for equivalency suggested by classmates that you and your partners didn't think of? If so, list them below.

b. Which do you think is the most scientifically sound test for equivalency? Which of the techniques do you prefer to use to create an equivalent force? Why do you prefer it? How do you know it is valid?

Combining Equivalent Forces

Since the goal in this section is to learn produce and measure forces of arbitrary strengths, you can start by combining equivalent forces to create a force scale.

5.3.3. Activity: Creating *F*, 2*F*, 3*F*, . . .

Describe how you could combine your equivalent forces to create forces of different strengths.

5.4 USING STANDARD UNITS TO MEASURE FORCE

So far you have been measuring forces in your own units using procedures that you and your partners have defined. If you measure forces and want to have scientists in another location understand your results, it would be convenient if everyone used the same force unit. The accepted standard unit for force is the *Newton*. The Newton is defined as the force that is needed to give a 1 kilogram mass an acceleration in which the velocity of an object increases by 1 meter per second each second. We're getting ahead of ourselves because we haven't defined the kilogram as a unit of mass yet. We will soon.

A scientifically rigorous way to measure a force of 1.0 Newtons is to take a 1.0-kg object that can move without friction and apply just the right force to it to get an acceleration of 1.0 m/s/s. That force would, by definition, be one Newton—at least to two significant figures. But it is a pain to have to go to all that trouble, and you can use a standard device for measuring force in Newtons instead.

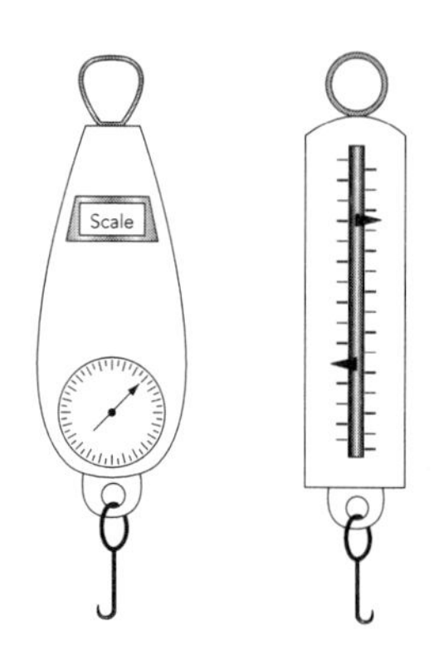

Fig. 5.5. Two types of spring scales used to measure forces in Newtons.

The most common device for measuring forces in Newtons consists of a spring with a scale attached to it that is marked off in Newtons. In theory, someone already figured out how much the spring had to stretch to get a 1-kilogram mass going at an acceleration of 1 m/s/s and combined forces to define the appropriate scale.

Another less common but very useful way to measure force is to use an electronic force sensor attached to a computer-based laboratory system that has been *calibrated* to read in Newtons.

In the activities in this section, you will measure forces in Newtons with both a spring scale and an electronic force sensor using the following equipment:

- 1 ruler
- 6 rubber bands, #14
- 1 spring scale, 10 N
- 1 computer-based laboratory system
- 1 force sensor
- 1 motion software

Recommended group size:	2	Interactive demo OK?:	N

Measuring Force in Newtons with a Spring Scale

You should devise a way to use rubber bands to show that the forces indicated on a spring scale in Newtons are proportional to your rubber band units.

5.4.1. Activity: Converting Your Units to Newtons

a. Devise a way to show that the forces indicated by a spring scale in Newtons are proportional to your rubber-band units. Explain what you did and show your data and graph in the space below.

b. Find a conversion factor between your personal rubber-band units and Newtons. Explain how you determined the factor and show any calculations in the space below.

Calibration of an Electronic Force Sensor

It is very useful to be able to read forces in Newtons using an electronic force sensor. Some types of electronic force sensors require calibration before they can record forces in Newtons. In general, calibration involves finding a relationship you or a computer program can use to convert readings on the measuring instrument to the quantity you want to measure. Procedures for force sensor calibration are either built into the motion software or require a mechanical adjustment in the sensor. To calibrate your force sensor using the spring scale as a standard:

1. Set up your computer-based laboratory system with a force sensor, motion detector, and interface.
2. Calibrate the force sensor. (If necessary, refer to the manual for the force sensor and associated software for calibration instructions.)

Note: If you are using a PASCO Force Sensor (Model CI–6537 or CI–6618), you will probably not need to calibrate it. In that case you might be asked to complete parts b. and c. of Activity 5.4.2 to confirm that the calibration is accurate.

5.4.2. Activity: Calibrating the Force Sensor in Newtons Using a Spring Scale

a. Use a 5-N force as indicated on the spring scale to calibrate your force sensor. What do you think are the major sources of uncertainty in your procedure?

b. To check the accuracy of your calibration, pull on the force sensor with a spring scale that reads 3 N of force. Is your calibration accurate? If not, repeat part a.

c. Select a force-time graph. Push and pull on the force sensor and look at the reading on your graph. Is one Newton a very large force? Explain.

d. What is the largest pulling (or positive) force you can measure before the force-time graph flattens out, indicating that the force sensor-computer system has been driven beyond its limits?

RELATING ACCELERATION AND FORCE

5.5 MEASURING ACCELERATION AS A FUNCTION OF FORCE

You have already determined that, for a situation in which friction is small, a constant force on an object causes it to accelerate at a constant rate. Now that you have a more thorough understanding of how forces can be defined and measured, you are ready to investigate the relationship between the magnitude and direction of the applied force and magnitude and direction of the resulting acceleration for different forces on a moment-by-moment basis.

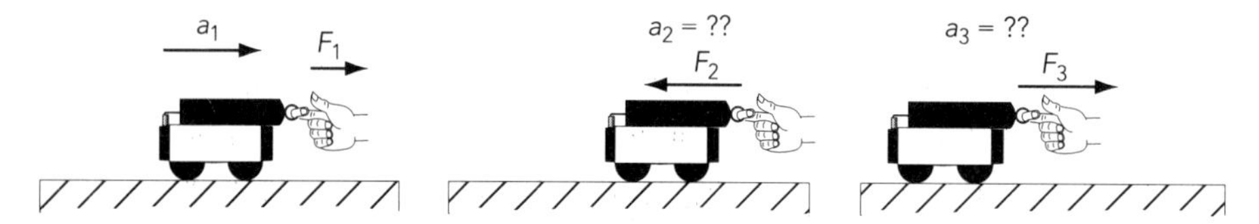

Fig. 5.6. How does the measured acceleration of a low-friction cart change when the applied force on it is changed?

5.5.1. Activity: Predicting Acceleration or Velocity vs. Force

a. Suppose you push and pull on a force sensor attached firmly to a low-friction cart and obtain a graph of force vs. time like that shown below. Do you expect the velocity vs. time or the acceleration vs. time graph to have the same shape as the force vs. time graph does? Explain the reason for your prediction in light of observations you have already made when applying a constant force to a person on a cart.

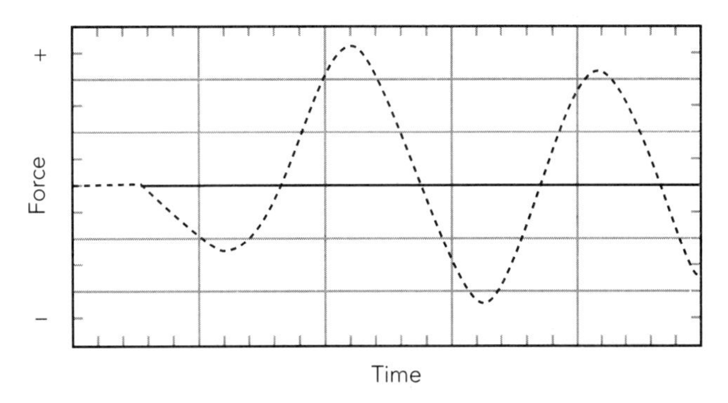

Fig. 5.7.

b. Consider the previous graph that shows sample force vs. time data. Sketch the predicted shape of the graph of corresponding velocity vs. time or acceleration vs. time. Please label the graph's vertical axis.

v or a?

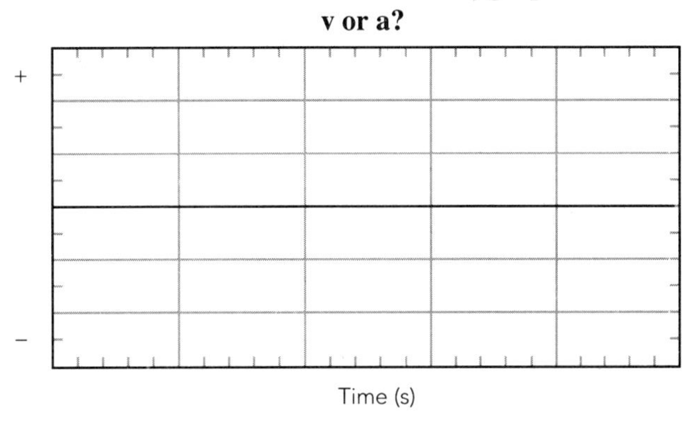

In order to investigate how acceleration and force are related, you will push and pull on a force sensor that is attached to a cart and record the motion of the cart. You will need:

- 1 small low-friction dynamics cart
- 2 masses, 500 g (to add mass to the cart)
- 1 smooth ramp or level surface 1–3 meters long
- 1 computer-based laboratory system
- 1 motion software
- 1 ultrasonic motion sensor
- 1 force sensor* (with a hook on its sensitive end)
- 1 adapter bracket (to attach a force sensor to the cart)
- 1 spring scale, 10 N (to calibrate the force sensor)

Recommended group size:	2	Interactive demo OK?:	Y

* Because of its linearity, low noise, and built in calibration, we recommend that the PASCO Force Sensor (models CI-6537 or CI-6618) be used for this activity.

To do the next activity you want to be able to push and pull on a cart and measure both the force and motion continuously. You can start by attaching a force sensor (with a hook on its end) firmly to a cart. Then you should add about 1.0 kg of mass to the cart and place it on a smooth level track or surface.

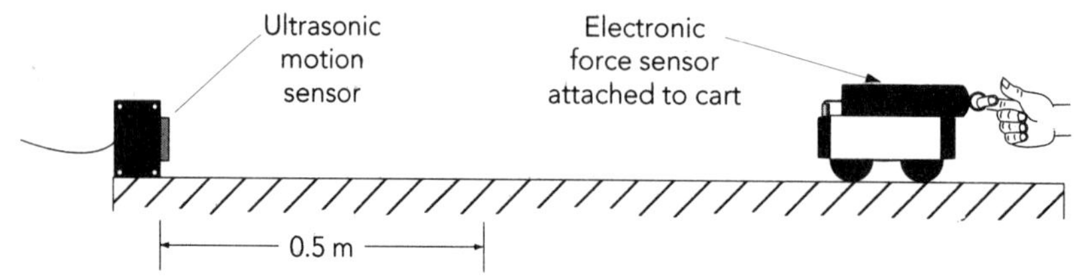

Fig. 5.8. Setup showing a motion sensor tracking the acceleration of a cart rolling on a level track as a force sensor detects the pushes and pulls on it. As usual, the cart must be at least 0.5 m away from the motion detector at all times.

If needed, calibrate the force sensor to read forces between 0 and 5 Newtons. Set up the motion software to display two graphs: Force vs. time and acceleration vs. time for about 5 seconds.

5.5.2. Activity: Measuring Acceleration vs. Force

a. Zero the force sensor and record data as you grip the hook on the end of the force sensor firmly to push and pull the cart back and forth on the track *smoothly*. Repeat this process until you have smooth reliable data. Then sketch the force and acceleration graphs. **Note:** If your acceleration graph seems rough and has spikes on it, you should set the averaging for the velocity and acceleration data at something between 5 and 15 points so that small uncertainties in data do not appear on the acceleration graphs.

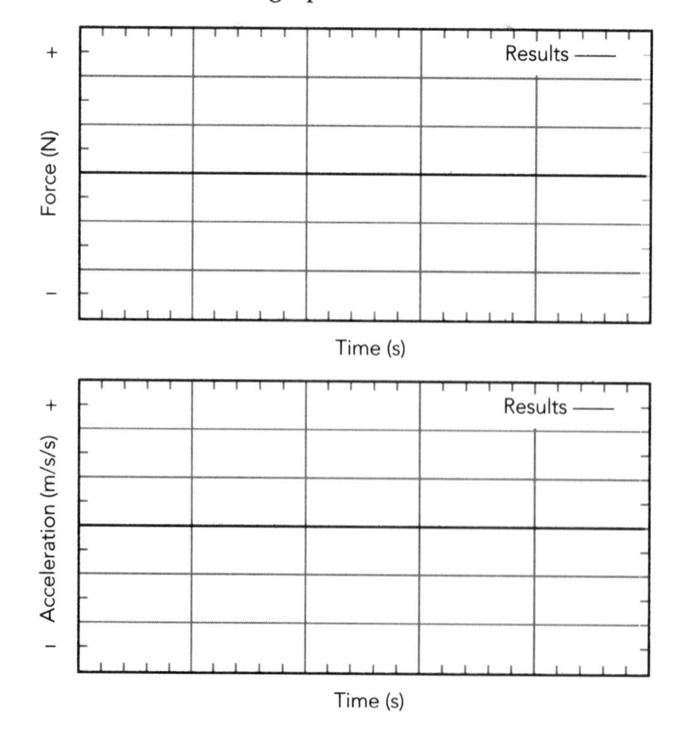

b. You should have observed that the force and acceleration graphs do indeed have the same shape. Is this what you predicted in Activity 5.5.1?

Are Force and Acceleration Proportional to Each Other?

The fact that the graphs of force and the resulting acceleration caused by it have the same general shape suggests that force and acceleration might be proportional to each other. Recall that if two variables are proportional, a graph of one variable as a function of the other is a straight line passing through the origin, and that the equation relating them would have the form

$$F = ma \tag{5.1}$$

where F represents the force, a represents the acceleration, and m represents the slope of the graph and is the constant of proportionality. Since you have a computer-generated table of values of a and F for a whole series of times, you can use the data you just gathered to test for proportionality between a and F.

5.5.3. Activity: Finding the Mathematical Relationship Between Acceleration and Force

a. Use the analysis feature of your motion software or display a data table to obtain about five sets of a and F values. Two significant figures will be fine. **Hint:** Find the corresponding values of F when a is the most negative, approximately zero, and the most positive. Then find a couple of sets of a and F values when a is between the most negative and zero and between the most positive and zero.

	a(m/s/s)	F(N)
1		
2		
3		
4		
5		

b. Create a graph of F vs. a with properly labeled axes including units and affix it on the following page. **Note:** It is conventional to plot the independent variable on the x-axis and the dependent variable on the y-axis. Using this convention, you should graph a as a function of F *because* F is the independent variable that you can change at will as you push or pull on the cart. However, it is more convenient in later activities if you graph F as a function a instead.

(Affix your graph here.)

c. Graphs of experimentally determined relationships are seldom perfect since there is usually some scatter due to uncertainties. Taking this fact into account, does the relationship between F and a appear to be a proportional relationship? Why or why not?

d. If the relationship is proportional, what is the constant of proportionality (i.e., the slope of the graph)? Explain how you determined the slope and include units for it. (Fitting, mathematical modeling, drawing a best slope by hand and estimating its value, etc.)

e. Write the general equation that relates F and a in terms of the symbol, m, which represents the slope of the graph.

You have just discovered a one-dimensional law of proportionality between an applied force and acceleration for the situation in which there is almost no friction. This is almost, but not quite,one of Newton's famous laws of motion. Before you can enrich the law of proportionality, you will need to explore how the properties of the object being accelerated affect the proportionality constant. In addition, you need to learn about what happens when more than one force acts on an object at the same time.

5.6 NET FORCE: ADDING AND SUBTRACTING FORCES

Fig. 5.9. What is the net effect of forces acting in opposite directions?

We have not yet thought formally about how forces combine. In doing so, it is useful to treat forces as mathematical entities. Let's postulate that force behaves mathematically like a *vector* quantity.

Quantities that have vector behavior are often denoted by a letter with a little arrow above it such as $\vec{F}$ or by a boldface letter such as **F**. The sum of several vectors is often denoted by placing a summation sign in front of a vector symbol to get $\Sigma\vec{F}$.

A one-dimensional vector is a mathematical entity that has both a direction along an axis and a magnitude. Vectors can have a direction along the positive x-axis or a direction along the negative x-axis. The magnitudes of one-dimensional forces can be represented by a single number, while the directions can be expressed in terms of a plus or minus sign in front of a unit vector. A one-dimensional vector can be represented as the product of its magnitude and direction as shown below.

1D forces acting in the same direction

$$\vec{F}_1 = +5.5$$
$$\vec{F}_2 = +3.4$$

1D forces acting in opposite directions

$$\vec{F}_1 = +2.3$$
$$\vec{F}_2 = -7.7$$

Reminder: $\vec{x}$ and $\vec{y}$ represent unit vectors pointing along the x- and y-axes, respectively. Many texts use $\hat{\imath}$ and $\hat{\jmath}$ for these quantities.

You can do some simple observations to determine whether or not one-dimensional forces behave like vectors. To do this you will need:

- 3 identical spring scales, 10 N
- 1 small low-friction cart

Recommended group size:	2	Interactive demo OK?:	N

5.6.1. Activity: Do 1D Forces Behave Like Vectors?

a. Describe what happens when a spring scale is hooked to one end of the cart and extended in a horizontal direction so that its force is equal to 2.0 N in magnitude. Does the cart move? If so, how? This should be a casual observation—no need to take any data.

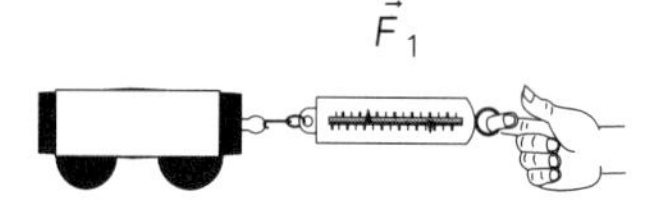

Fig. 5.10.

b. Draw an arrow that represents a scale drawing of the magnitude and direction of the force you are applying. Let one centimeter of arrow length represent each Newton of force. Label the arrow with an $\vec{F}_1$.

c. Observe what kind of motion results when two spring scales are hooked to opposite ends of the cart and extended in a horizontal direction so that each of their forces is equal to 2.0 N in magnitude but opposite in direction. Does the cart move? If so, how. What is the combined or net applied force on the cart?

Fig. 5.11.

d. Draw arrows that represent a scale drawing of the magnitudes and directions of the forces you are applying. Let one centimeter of arrow length represent each Newton of force. Label each arrow appropriately with an $\vec{F}_1$ or an $\vec{F}_2$.

e. What kind of motion results when two identical springs are displaced by the same amount in the same direction (e.g., when each spring is displaced to give 1.0 N of force)? How does this compare to the force of one spring displaced by twice that amount (e.g., so that it can apply 2.0 N of force)? Describe what you did and the outcome.

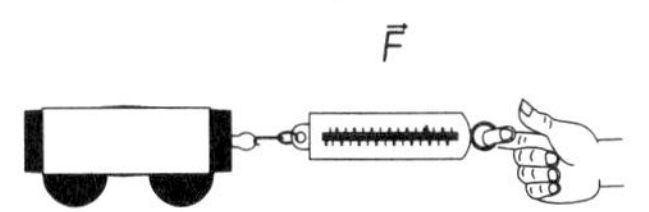

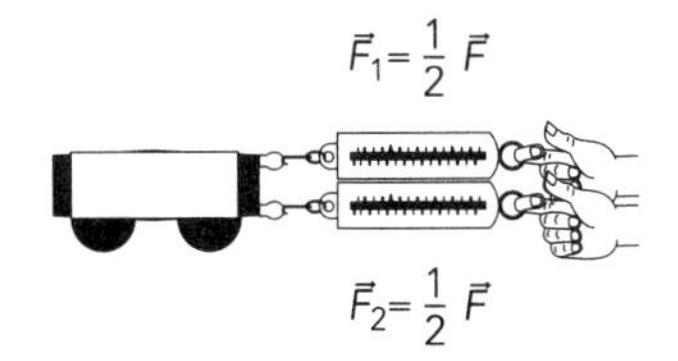

Fig. 5.12.

f. Draw arrows that represents a scale drawing of the magnitudes and directions of $\vec{F}_1$, $\vec{F}_2$, and $\vec{F}$.

g. Do one-dimensional forces seem to behave like one-dimensional vectors? Why or why not?

If forces can be described as vectors, then we can denote combinations of vectors such as those in Activity 5.6.1c and e by the equation

$$\sum \vec{F} = \vec{F}_1 + \vec{F}_2$$

where $\Sigma\vec{F}$ represents the vector sum of two or more forces. Some textbooks refer to a combined force as a net force. Other text authors write about the resultant force. Combined, resultant, or *net* force all refer to the same thing.

5.6.2. Activity: Calculating a Net Force

a. Suppose a first force has a magnitude of $F_1 = 1.5$ N and acts from right to left on a cart and a second force has a magnitude of $F_2 = 0.9$ N and acts from left to right on a cart. Using an x-axis, with the positive direction being from left to right, express F_1 and F_2 in vector notation. Recall that $\vec{x}$ is a unit vector pointing in the positive x-direction.

b. Express the net force in proper vector notation and explain how you calculated it.

You should note that velocities and accelerations can also be described by vectors because they too have magnitudes *and* directions.

One of your goals is to continue to refine your understanding of the relationship between one-dimensional forces and motion. You just investigated how one-dimensional forces combine like vectors in terms of their ability to cause acceleration. *We can now state that acceleration caused by several forces acting in one dimension on an object that experiences very little friction is proportional to the net force on the object.*

5.7 WHAT HAPPENS WHEN THE NET FORCE IS ZERO?

Let's consider an important special situation in more detail—the situation in which the net force acting in one dimension on an object in the absence of significant friction is zero. You can start by summarizing what you already know.

5.7.1. Activity: Facts About Zero Net Force and Motion

a. Suppose we apply a force in the positive direction (left the right) on a small low-friction cart and another force acting along the same line of equal magnitude in the negative direction on the cart (right to left). If the law of proportionality between force and acceleration holds, what will the acceleration of the cart be?

b. What is the acceleration of a cart that is:

1. at rest?

2. moving with a constant negative velocity?

3. moving with a constant positive velocity?

c. Suppose an object is moving with a constant velocity and experiences no net applied forces and no friction. Is its continued motion with a constant velocity compatible with the law of proportionality between net force and acceleration or does it violate that law?

d. What can you say about the net force on the cart mentioned in part b for each of the three cases?

We would like you to investigate whether or not a low-friction cart that has a zero net force on it can move at a constant velocity. To undertake this investigation we suggest that you apply forces in opposite directions on a cart using two hanging weights as shown in Figure 5.13.

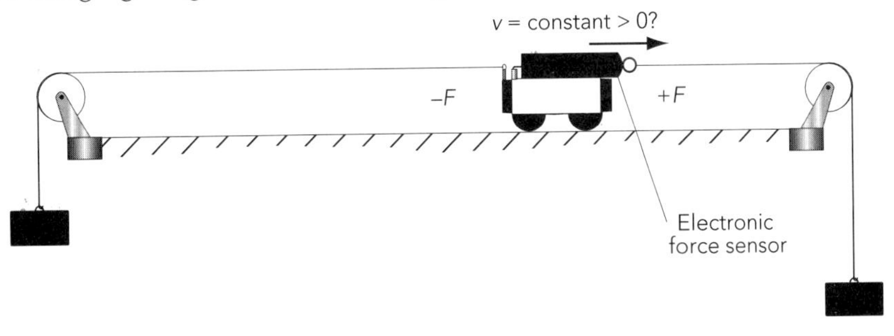

Fig. 5.13. Setup showing a low-friction cart rolling on a level track as two hanging weights combine to exert zero net force on the cart.

We recommend that this activity be done as a demonstration with the entire class participating. The items needed for the demonstration include:

- 1 high table
- 1 smooth ramp or level surface 2 meters long
- 1 small low friction dynamics cart
- 2 lengths of string, 2 m
- 2 low friction pulleys
- 2 mass pans, 1 kg
- 1 set of slotted masses (5 g to 100 g)
- 1 computer-based laboratory system

- 1 force sensor
- 1 motion software

Recommended group size:	All	Interactive demo OK?:	Y

Before making the observations you should level the track and balance the masses on either side with the smaller slotted masses so that, if the cart is stationary at first, it stays at rest and doesn't accelerate in either direction. An electronic force sensor should be firmly attached to the cart to help verify that the pulling forces on the cart don't change as the cart moves to the right or the left during the observations.

5.7.2. Activity: Motion with No Net Force

a. Suppose a cart receives a brief push that starts it moving in one direction or another. If there is no net force on it after the push, what do you predict its motion will be like? Try to imagine that there is no friction acting on the cart. Explain the reasons for your prediction.

b. Is your prediction compatible with the proportional relationship between force and acceleration that you discovered previously?

c. Observe what happens after the cart is pushed in one direction and allowed to move freely with no net force. Describe your observation. Is the velocity of the cart constant or decreasing. Does the cart seem to be accelerating?

d. The observation you just made should enable you to state Newton's First Law of Motion. *Please finish the statement in a way that is compatible with your actual observation.*

NEWTON'S FIRST LAW: If an object moving at a constant velocity, $\vec{v}$, without friction experiences no net force, it will . . .

Remarks About Newton's First Law

Newton's First Law is of deep significance because it allows an observer who is moving at a constant velocity with respect to another observer to discover the same laws of motion. For example, suppose you observe that the small cart is at rest in the laboratory and your partner makes the same observation while moving away from you at a constant velocity on a Kinesthetic cart. Your partner will see the small cart moving away with a constant velocity. But both of you can agree that the cart is not accelerating and therefore is experiencing no net force!

RELATING FORCE, MASS, AND ACCELERATION

5.8 DEFINING AND MEASURING MASS

Fig. 5.14. Would you rather push a sports car or a larger van?

A good friend calls you in a panic. His battery is dead and he needs to have you come outside and help push his car to get it started. He needs to have you get it moving at about 12 mph so he can throw it in gear and turn over the engine. You blithely answer sure, you'll be right out. Then you remember that your friend owns two vehicles—a large delivery van and a smaller sports car. Since you can exert only so much force and you're feeling like an 80 lb weakling, you hope that your friend is driving the easiest of the two cars to push.

5.8.1. Activity: Causing a Car to Accelerate

a. Which vehicle do you think would be easier to accelerate from rest to a speed of 12 mph with a given force—a small car or a large van? Explain.

b. What characteristic of an object seems to determine how much force is needed to accelerate it?

Somehow the magnitude of force required to cause an object to accelerate by a given amount is related to the "amount of stuff" being accelerated. It is pretty obvious to most people that if there is more stuff, then more force will be required to accelerate it. But suppose we double the amount of stuff. Will that mean that twice the force is needed to accelerate double the stuff?

The stuff we are referring to is what scientists usually call mass. Let's take some time to consider the question of what mass is and how we might measure it.

What Is Mass?

Philosophers of science are known to have great debates about the definition of mass. If we assume that mass refers somehow to "amount of stuff," then we can develop an operational definition of mass for matter that is made up of particles that appear to be identical. We can assume that mass adds up and that two identical particles when combined have twice the mass of one particle; three particles have three times the mass; and so on. But suppose we have two objects that have different shapes and are made of different stuff, such as a small lead pellet and a silver coin. How can we tell if these two entities have the same mass?

5.8.2. Activity: Ideas About Mass and Its Measurement

a. Attempt to define *mass* in your own words without using the word "stuff."

b. How many different ways can you think of to determine whether a lead pellet and a silver coin have the same mass?

c. Suppose you find that the lead pellet and the silver coin seem to have the same mass. How could you create "stuff" that has twice the mass of either of the original objects?

Using a Mass Balance

One time-honored way that people have used to compare the mass of two objects is to put them on a balance. If they happen to balance each other, we say that the "force of gravity" or the force of attraction exerted on them by the earth is the same, so they must have the same mass.

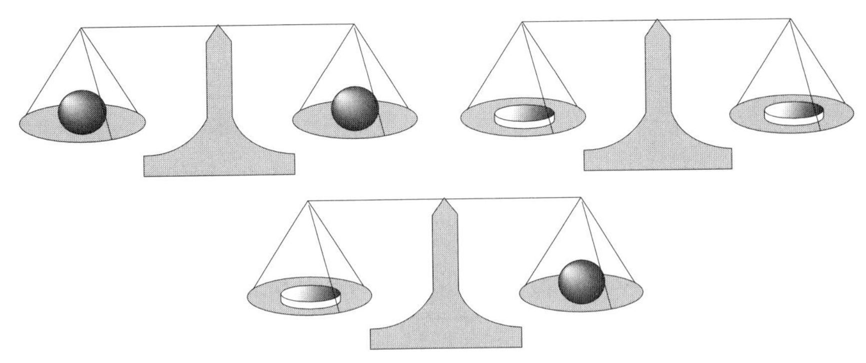

Fig. 5.15. A common method of determining mass that assumes two objects have the same mass if they experience the same gravitational force.

Fig. 5.16.

Actually, if we balance gravitational forces as the method of determining mass, we are only determining a *passive gravitational mass.* A passive gravitational mass is proportional to the force of attraction exerted by the earth on the mass. How can we use a balance to measure the mass of any object relative to a standard mass?

Let's do a thought experiment. Suppose Maya makes the outrageous claim that her dove has the same mass as 79 silver quarters and is worth her weight in quarters. Can you doublecheck her claim using a balance, a quarter as the "standard mass," and a pile of sand? **Hint:** This is an exercise in basic logic.

5.8.3. Activity: Using a Balance to Measure an Arbitrary Mass

Explain how you might measure the passive gravitational mass of Maya's dove using the balance, sand, and standard coin.

Other Ways to Measure Passive Gravitational Mass

In most modern laboratories, spring scales and electronic scales that are easier to use now replace the old-fashioned balance. As the earth attracts a mass hanging from a spring, the spring will stretch. A mass placed on the platform of an electronic scale will cause it to depress. The amount of depression can be detected electronically.

5.9 HOW MASS AFFECTS MOTION

You have already verified a law of proportionality between force and acceleration when little friction is present. This law can be expressed in the form

$$F = ma \tag{5.1}$$

where F is the force exerted on an object, m is the slope of the graph of F vs. a or the constant of proportionality, and a is the acceleration caused by the force. We know that this constant of proportionality, m, which represents a resistance of an object to acceleration, doesn't necessarily have anything to do with gravity.

Since it requires more force to accelerate more gravitational mass our intuition tells us that the proportionality constant ought to be related to passive gravitational mass. Is it possible that the passive gravitational mass of an object is the same as the proportionality constant, or slope, relating F and a? In other words, *will accelerating twice as much passive gravitational mass by the same amount take twice as much force?* This is the question we posed in Section 5.7.

To answer this question you should investigate the forces and accelerations that arise when you push and pull on rolling carts having different passive gravitational masses. This investigation is basically an extension of the one you undertook in Section 5.5. For the next activity you will need:

- 1 balance
- 2 pieces of string (to use with the balance)
- 1 electronic scale
- 1 small low-friction dynamics cart
- 1 set of assorted masses (1g, 2 g, 5 g, 10 g, 20 g, 50 g, 100 g, 200 g)
- 1 smooth ramp or level surface 1–3 meters long
- 1 computer-based laboratory system
- 1 motion software
- 1 ultrasonic motion sensor
- 1 force sensor*
- 1 adapter bracket (to attach a force sensor to the cart)
- 1 spring scale, 10 N (to calibrate the force sensor if needed)

Recommended group size:	2	Interactive demo OK?:	Y

Before doing the activity, attach a force sensor with a hook on its end firmly to the cart. Place the cart on a smooth level track or surface.

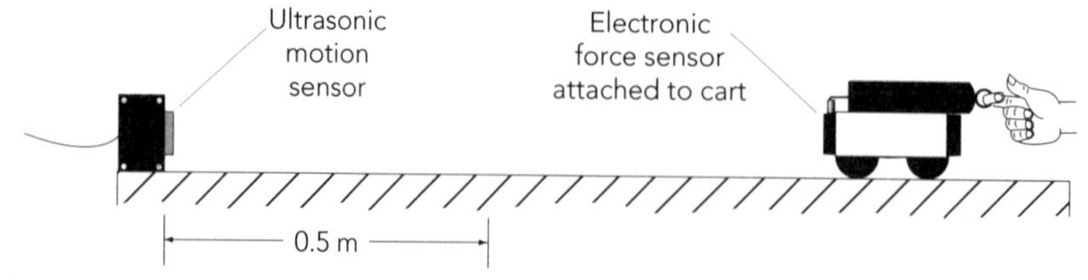

Fig. 5.17. Setup showing a motion sensor tracking the acceleration of a cart rolling on a level track as a force sensor detects the pushes and pulls on it. As usual, the cart must be at least 0.5 m away from the motion detector at all times.

When you did a similar observation in Section 5.5 you collected data for

* Because of its linearity, low noise, and built-in calibration, we recommend that the PASCO Force Sensor (models CI-6537 or CI-6618) be used in this activity.

force vs. time and acceleration vs. time to see the shapes of the graphs. Since you were also interested in the relationship between force and acceleration you plotted a graph of this relationship using some sample data and found the slope of the graph. This time you should set up the motion software to display three graphs:

1. force vs. time
2. acceleration vs. time
3. force vs. acceleration

If needed, calibrate the force sensor to read forces between -5 and $+5$ N.

5.9.1. Activity: Measuring Acceleration vs. Force

a. Zero the force sensor (if needed) and record data as you grip the force sensor hook firmly to push and pull the cart back and forth on the track smoothly. Sketch the force and acceleration graphs you observed. **Note:** If your acceleration graph seems rough and has spikes on it, (1) you should set the averaging for the velocity and acceleration data at something between 5 and 15 points so that small uncertainties in data do not appear on the acceleration graphs, and (2) you should try to push and pull with a maximum force of about 4 N.

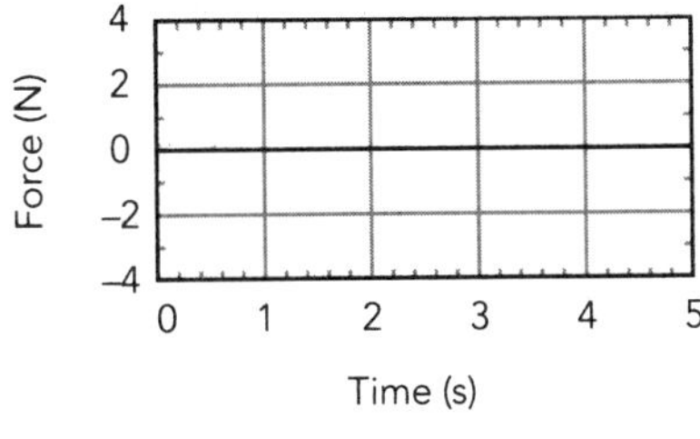

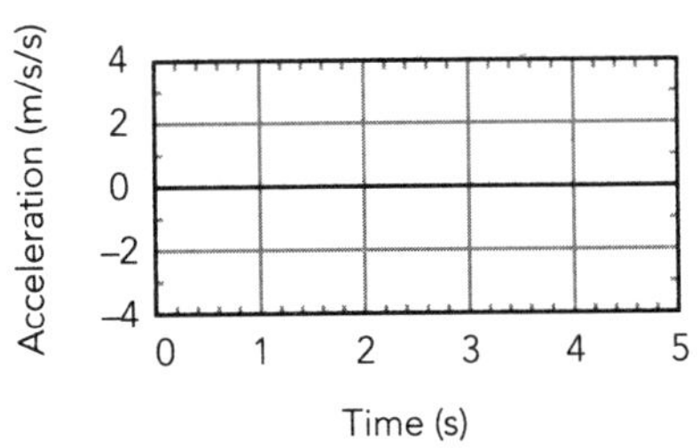

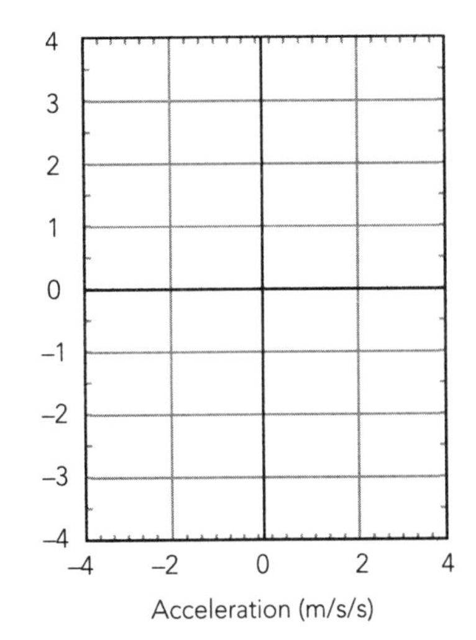

b. Use the linear fitting feature in the motion software to find the slope of the graph of force vs. acceleration. What is the value of the slope and its standard error? Be sure to include units.

Slope: m =

Standard Deviation from the Mean: SDM =

Doubling the Gravitational Mass

Suppose you put enough mass on the cart to double its mass. How much more force must you exert to push and pull on the more massive cart so that its acceleration vs. time graph looks about the same as the graph you obtained in Activity 5.9.1a?

5.9.2. Activity: Acceleration vs. Force for Twice the Mass

a. Suppose you were able to push and pull on a cart having twice the mass as before so that the acceleration vs. time is approximately the same. Do you think the slope of the F vs. a graph will increase, decrease, or remain the same? Explain.

b. Sketch the predicted shapes of all three graphs.

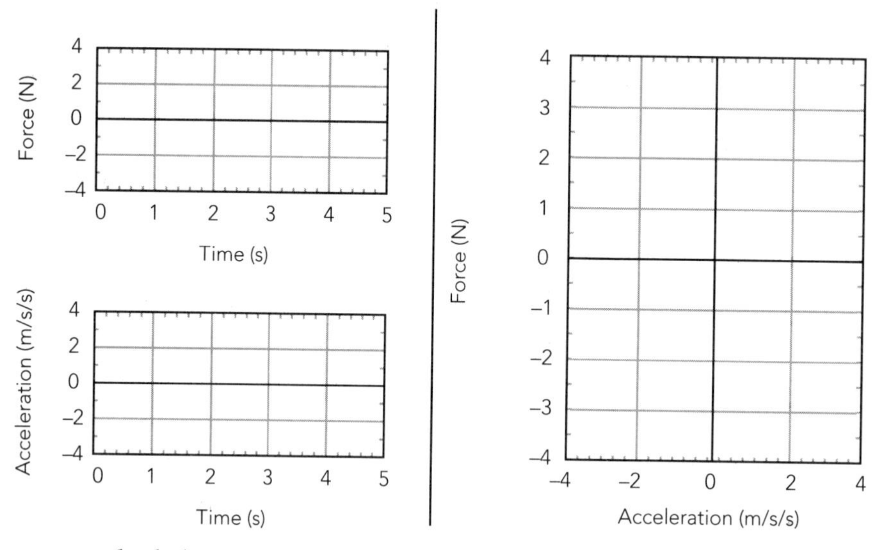

c. Use the balance to compare the combined mass of the cart, force sensor, and sensor holder to that of a collection of assorted masses. Include enough assorted masses to just balance the gravitational mass of the cart-sensor system.

Place this combination of masses on the cart so that the cart with added masses, force sensor, and sensor holder now have twice the gravitational mass as before. Zero the force sensor (if needed) and record data as you grip the force sensor hook firmly to push and pull the cart back and forth on the track *smoothly*. Sketch the force and acceleration graphs you observed.

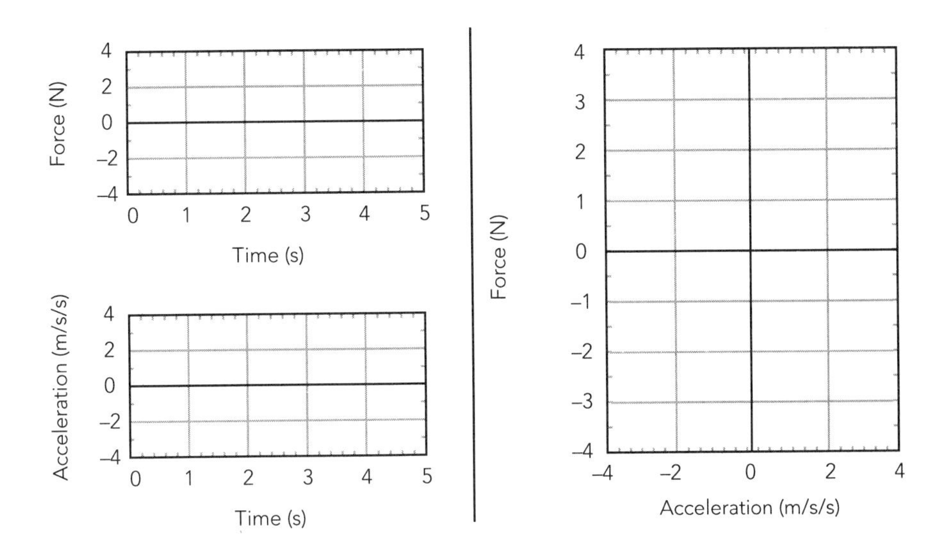

d. Using the linear fit feature in the motion software, indicate the value of the slope and the standard error of the force vs. acceleration graph. Include units for the slope.

Slope: m =

Standard Deviation from the Mean: SDM =

e. Compare the slope you found here with that found in Activity 5.9.1. Does gravitational mass seem to be the constant of proportionality between force and acceleration?

Newton's Second Law

The slope of the linear graph relating force and the acceleration caused by that force is defined as the *inertial mass.* By defining inertial mass in this way, we have developed sensible definitions of force and mass that lead to Newton's Second Law of Motion for one-dimensional motions in the absence of friction. Newton's Second Law expressed in vector notation for an object having a total inertial mass of m is simply

$$\sum \vec{F} = m\vec{a} \tag{5.2}$$

where $\Sigma\vec{F}$ is a vector representing the net force on the object, and $\vec{a}$ is a vector representing the acceleration caused by the net force.

Inertial and Passive Gravitational Mass

In the last activity you have shown that within the limits of experimental uncertainty inertial and passive gravitational masses are the same. This makes the definition of inertial mass as the proportionality constant between acceleration and force seem less arbitrary.

It is not obvious that these two definitions of mass—passive gravitational and inertial—should yield exactly the same results. This equivalence is assumed in both Newton's theory of gravity and Einstein's general relativistic modifications of it. In fact, sophisticated experiments have shown that within the limits of experimental uncertainty, there is no difference between the two types of mass to within one part in 1011.

Standard SI Units for Mass and Force

As you already know, the Systemè Internationale, or SI, system of units was established in 1960 to provide standard units for all scientists to use throughout the world. The SI units for fundamental quantities used in mechanics, including mass, are shown below.

SI UNITS FOR MECHANICS

Length: A **meter** (m) is the distance traveled by light in a vacuum during a time of 1/299,792,458 second.

Time: A **second** (s) is defined as the time required for a light wave given off by a cesium–133 atom to undergo 9,192,631,770 vibrations.

Mass: A **kilogram** (kg) is defined as the mass of a platinum–iridium alloy cylinder kept in a special chamber at the International Bureau of Weights and Measures in Sévres, France.

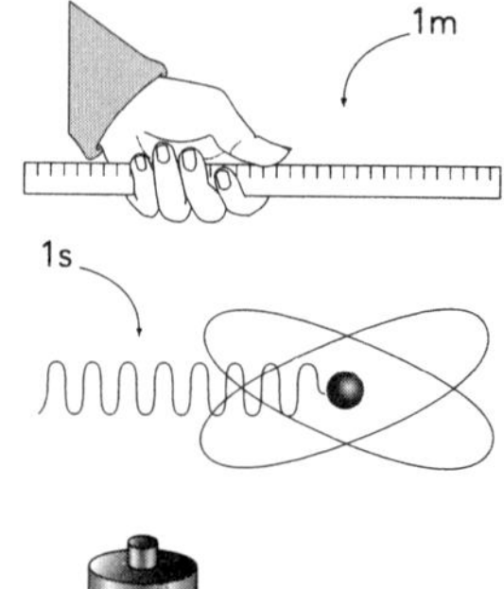

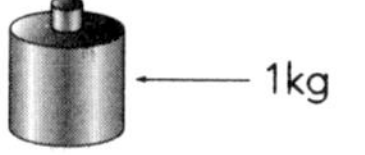

Fig. 5.18.

The electronic balance and spring scales often used in laboratories have been calibrated using replicas of the "real" standard kilogram mass kept in a vault in France. These fundamental units and Newton's Second Law can also be used to define the Newton as a unit of force. The Newton is defined in terms of mass and acceleration as shown in the box below.

> *The Force Unit Expressed in Terms of Length, Mass, and Time Force:* A **Newton** (N) is defined as that force which, when acting on a 1-kg mass, causes an acceleration of 1 m/s/s.

Fig. 5.19.
1N = 1 kg • m/s/s

Do the Standard Units Work Together?

With the exception of a device called an inertial balance, essentially all the common equipment used in laboratories today measures passive gravitational mass rather than inertial mass. Thus, if you determine the mass in kilograms of your cart and force sensor system using an electronic balance, you can compare it to the inertial mass you found by accelerating your cart system in Activity 5.9.1.

Since you measured forces in Newtons and accelerations in m/s/s, the passive gravitational mass readings from a well-calibrated electronic balance and the inertial mass readings from the slopes of your F vs. a lines should be the same. Are they?

5.9.3. Activity: Gravitational vs. Inertial Masses in Kilograms

a. Use the electronic balance to determine the passive gravitational mass of your cart with the force sensor attached to it.

$m_{grav} =$ kg

b. What is the inertial mass of your cart with the force sensor attached to it as reported in Activity 5.9.1?

$m_{inertial} =$ kg

c. Are they the same within the limits of experimental uncertainty?

d. What do you think are the possible sources of systematic error and experimental uncertainty in your two measurements?

5.10 SUMMARIZING NEWTON'S FIRST AND SECOND LAWS

The main purpose of Unit 5 has been to explore the relationships between forces on an object, its mass, and its acceleration. You have been trying to develop Newton's first two laws of motion for one-dimensional situations in which all forces lie in a positive or negative direction along the same line and in which there is very little friction present.

5.10.1. Activity: Newton's Laws in Your Own Words

Express Newton's Laws in your own words clearly and precisely.

- The First Law (the one about constant velocity):

- The Second Law (the one relating force, mass, and acceleration):

Fig. 5.20.

5.10.2. Activity: Newton's Laws in Equation Form

Express Newton's Laws in equations in terms of the acceleration or velocity vector, the net force on an object, and its mass:

The First Law: If $\sum \vec{F} = 0$ then $\vec{v} =$ or

Note: The use of the equal sign does not signify that an acceleration is the same as or equivalent to a force divided by a mass, but instead it spells out a procedure for calculating the magnitude and direction of the acceleration of a mass while it is experiencing a net force. What we assume when we believe in Newton's Second Law is that a net force on a mass *causes* an acceleration of that mass.

The Second Law: If $\sum \vec{F} \neq 0$ then $\vec{a} =$

Fig. 5.21. Newton recognized that the laws of motion discovered through the use of applied forces could also be used to discover the nature of gravitational forces and the forces of friction.

Final Comments on Force, Mass, and Motion

You started your study of Newtonian dynamics in this unit by attempting to develop the concept of force. Initially, when asked to define force, most people think of a *force* as an *obvious push or pull* such as a punch to the jaw or the tug of a rubber band. By studying the acceleration that results from a force when little friction is present, we can come up with a second definition of *force* as *that which causes acceleration.* These two alternative definitions of force do not seem to be the same at all. Pulling on a hook attached to a wall doesn't seem to cause the wall to move. An object dropped close to the surface of the earth accelerates and yet there is no visible push or pull on it.

The genius of Newton was to recognize that he could define *net force* as that which causes acceleration. He reasoned that if the applied forces did not account for the degree of acceleration then other "invisible" forces must be present. A prime example of an invisible force is that of gravity—the attraction of the earth for objects.

Finding invisible forces is hard sometimes because some of them are known as *passive* forces because they only seem to act in response to either the motion of an object or other forces on it. Friction forces are one example of passive forces. They are not only invisible, but they only crop up during motions for the purpose of inhibiting the motion. The passive nature of friction is obvious when you think of a person riding on a garbage bag and sliding along the floor at constant velocity under the influence of an applied force.

According the Newton's First Law, a person moving at a constant velocity must have no net force on her. Newton thought that the applied force in one direction had to be opposed by a friction force acting in the other direction to oppose her motion. The friction force must be passive because, if the applied force is discontinued, the friction force does cause the sliding person to slow down until she has no motion. Then the friction force stops acting. If it didn't stop acting, the person would slow down and then turn around and speed up in the opposite direction. This doesn't happen!

During the rest of our study of the Newtonian formulation of classical mechanics your task will be to discover and invent new types of active and passive forces so that you can continue to explain and predict motions using Newton's Laws. In the next unit you will be using Newton's Laws to explain why sliding masses that have no visible forces on them slow down and come to rest, and to learn why masses fall when dropped close to the surface of the Earth.

Name ______________________ Section __________ Date __________

UNIT 6: GRAVITY AND PROJECTILE MOTION

Dolphins are powerful, graceful, and intelligent animals. As this dolphin leaps out of the water in delighted play, she experiences a change in velocity that is the same as that of any other mass moving freely close to the surface of the earth. She is undergoing what physicists call projectile motion. The path she follows while above the water has the same mathematical characteristics as the path of a ball, a ballet dancer performing a grand jeté, or Michael Jordan executing a slam dunk. What is this path like? What kind of force causes projectile motion? In this unit we will use a faith in Newton's Second Law to discover the nature of the Earth's gravitational attraction for masses and to explore the mathematical behavior of free motions close to the Earth's surface.

UNIT 6: GRAVITY AND PROJECTILE MOTION

Science is a game . . . with reality. . . . In the presentation of a scientific problem, the other player is the good Lord. He has . . . devised the rules of the game—but they are not completely known, half of them are left for you to discover or deduce . . . the uncertainty is how many of the rules God himself has permanently ordained, and how many apparently are caused by your own mental inertia, while the solution generally becomes possible only through freedom from its limitations. This is perhaps the most exciting thing in the game. For here you strive against the imaginary boundary between yourself and the Godhead—a boundary that perhaps does not exist.

Erwin Schrödinger

OBJECTIVES

1. To explore the phenomenon of gravity and study the nature of motion along a vertical line near the earth's surface.
2. To use Newton's laws to invent or discover invisible forces for describing phenomena such as "gravity."
3. To learn to describe positions, velocities, and accelerations using vectors.
4. To understand the experimental and theoretical basis for describing projectile motion as the superposition of two independent motions: (1) a body falling in the vertical direction, and (2) a body moving in the horizontal direction with no forces.
5. To observe the similarity between the type of motion that results from projectile motion and that which results from tapping a rolling ball continuously.

6.1. OVERVIEW

When an object is dropped close to the surface of the earth, there is no obvious force being applied to it. Whatever is causing it to pick up speed is invisible. Most people casually refer to the cause of falling motions as the action of "gravity." What is gravity? Can we describe its effects mathematically? Can Newton's laws be interpreted in such a way that they can be used for the mathematical prediction of motions that are influenced by gravity? We will study the phenomenon of gravity for vertical motion in the first few activities in this unit.

Next you will prepare for the mathematical description of two-dimensional motion by learning about some properties of two-dimensional vectors, that can be used to describe positions, velocities, and accelerations. Finally, you will study projectile motion, in which an object accelerates in one dimension and moves at a constant velocity in the other.

VERTICAL MOTION

6.2 DESCRIBING HOW OBJECTS FALL*

Let's begin the study of the phenomenon of gravity by predicting and then observing the motion of a couple of falling objects. For these activities you will need:

- 1 small rubber ball
- 1 flat-bottomed coffee filter

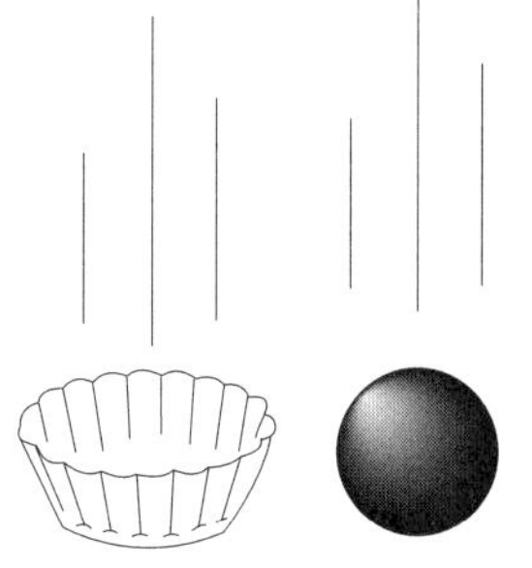

Fig. 6.1.

Recommended group size:	All	Interactive demo OK?:	Y

6.2.1. Activity: Predicting Falling Motions

a. Predict how the ball will fall in as much detail as possible. For example, will it speed up quickly to a constant falling velocity and then fall at that velocity? If not, what will it do? Explain the reasons for your prediction.

b. Predict how the coffee filter will fall in as much detail as possible. For example, will it speed up quickly to a constant velocity of fall? If not, what will it do? Do you think the filter will fall more slowly or more rapidly than the rubber ball? Explain the reasons for your prediction.

c. Now predict how the coffee filter will fall *if it is crumpled up into a little ball.* For example, will it speed up quickly to a constant velocity of fall? If not, what will it do? Do you think the crumpled filter will fall more slowly or more rapidly than the rubber ball? Explain the reasons for your prediction.

* This activity was suggested by Professor Richard Hake of the University of Indiana.

At this point you should observe what actually happens. Someone in your class should stand on a table or chair and drop the rubber ball and the coffee filter at the same time. Then the observation should be repeated with the rubber ball and the crumpled coffee filter falling at the same time.

6.2.2. Activity: Predicting Falling Motions

Describe your observations. How did they compare with your predictions. If your predictions and observations were not the same in each case, develop a new explanation for your observations.

6.3. DESCRIBING HOW OBJECTS RISE AND FALL

Let's continue the study of the phenomenon of gravity by predicting the nature of the motion of an object, such as a small rubber ball, when it is tossed up and then allowed to fall vertically near the surface of the earth. This is not easy since the motion happens pretty fast! To help you with this prediction, toss a ball in the laboratory several times and see what you think is going on. After making the prediction, you will be asked to use a technique for studying the motion in more detail than you can by using direct observation. You can either analyze a video of the motion of a ball toss or use a motion sensor with a computer-based laboratory system to record position vs. time for a toss. In order to make the prediction and observations, you will need:

- 1 small rubber ball
- 1 video analysis system w/VideoPoint software
- 1 meter stick

—or—

- 1 basketball
- 1 computer-based laboratory system
- 1 ultrasonic motion sensor
- 1 spreadsheet software

Recommended group size:	3	Interactive demo OK?:	Y

6.3.1. Activity: Predicting the Motion of a Tossed Ball

a. Toss a ball straight up a couple of times and then describe how you think it might be moving when it is moving *upward*. Some possibilities include: (1) rising at a constant velocity; (2) rising with an increasing acceleration; (3) rising with a decreasing acceleration; or (4) rising at a constant acceleration. What do you think?

b. Explain the basis for your prediction.

c. Now describe how you think the ball might be moving when it is moving *downward*. Some possibilities include: (1) falling at a constant velocity; (2) falling with an increasing acceleration; (3) falling with a decreasing acceleration; or (4) falling at a constant acceleration. What do you think?

d. Explain the basis for your prediction.

e. Do you expect the acceleration when the ball is rising to be different in some way than the acceleration when the ball is falling? Why or why not?

f. What do you think the acceleration will be at the moment when the ball is at its highest point? Why?

The motion of a tossed ball is too fast to observe carefully by eye without the aid of special instruments. We recommend that you use a video camera to film the ball at a rate of 30 frames per second; you can then replay the film a single frame at a time. **Note:** If you don't have a video analysis system available, you may analyze a ready-made digitized movie of a ball toss using video analysis software. Alternatively, you can use a computer-based laboratory system with a motion detector to track the motion of a basketball.

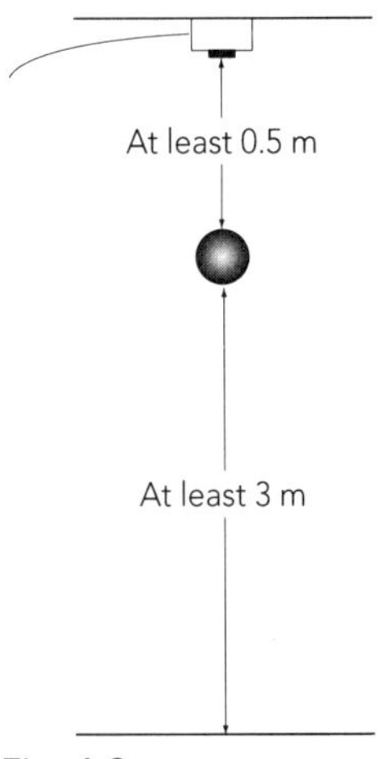

Fig. 6.2.

To Use a Computer-Based Lab

1. Tape the motion detector to a light fixture or something else as high above the floor as possible, with the detector looking downward, as shown on the right.
2. Since the fall is quite rapid, set up the motion software to record position vs. time at 30 points/second.
3. If you want to use a conventional coordinate system in which down is negative, you should set the motion software to designate distance from the motion sensor as negative.
4. It is the data describing a bounce that has the rise and fall in it. Thus, as soon as you start recording data, drop the basketball and allow it to bounce once. Transfer just the bounce data to your spreadsheet.

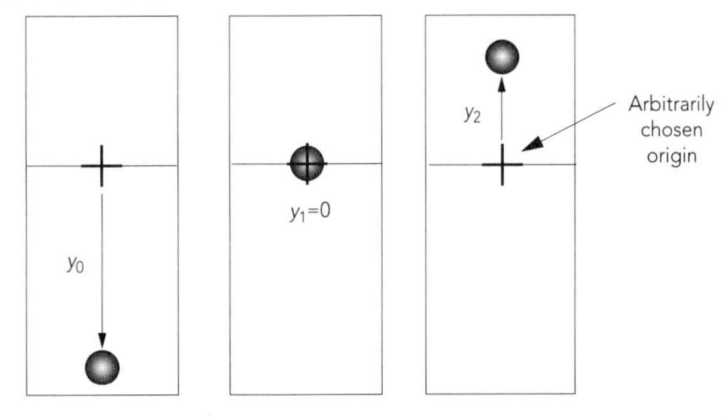

Fig. 6.3. Three "snapshots" of a ball rising showing suggested notation for keeping track of its positions as recorded by a video analysis system. Which y–values are positive and which are negative? The location of the origin is chosen arbitrarily.

Fig. 6.4.

6.3.2. Activity: Observing Motion of a Ball

a. Follow the directions to find times and corresponding distances from an origin of your choice as the object rises and falls. Record up to 36 data points in the table below or attach a computer printout of your data table.

	t(s)	y(m)
1.		
2.		
3.		
4.		
5.		
6.		
7.		
8.		
9.		
10.		
11.		
12.		

	t(s)	y(m)
13.		
14.		
15.		
16.		
17.		
18.		
19.		
20.		
21.		
22.		
23.		
24.		

	t(s)	y(m)
25.		
26.		
27.		
28.		
29.		
30.		
31.		
32.		
33.		
34.		
35.		
36.		

b. Place the data in a spreadsheet and use a computer graphing routine to plot a graph of y vs. t. Affix a copy of the graph in the space below.

c. Use a sketch to describe where you decided to place your origin. What is the initial value of y (usually denoted y_0) in the coordinate system you chose?

d. By examining your data table, calculate the *approximate* value of the initial velocity of the ball in the y-direction. Include the sign of the velocity and its units. (Use the convention that on the y-axis up is positive and down is negative.)

Now that you have collected and graphed data for the rise and fall of a ball, in the next two activities you will consider the types of motion separately for the rise and the fall. In particular, you will explore whether the rise and fall accelerations are the same or different.

6.3.3. Activity: How a Ball Actually Rises

a. Examine the portion of the graph in Activity 6.3.2b that represents the upward motion of the ball. What is the nature of this *upward* motion? Constant velocity, constant acceleration, an increasing or decreasing acceleration? How does your observation compare with the prediction you made in Activity 6.3.1?

b. Using the convention that on the y-axis up is positive and down is negative, is the *acceleration* of the tossed object positive or negative as it rises (i.e., in what direction is the magnitude of the velocity increasing)?

c. If you think the object is undergoing a *constant* acceleration, use the modeling technique you used in Unit 4, Activity 4.10.1, to find an equation that describes y as a function of t as the ball rises. **Hints:** You might try to model the system with kinematic equation number 1. Since you are dealing with the second dimension (i.e., the vertical dimension), you should replace x with y and x_0 with y_0. Write the equation of motion in the space below. Then use coefficients of time to find the values of a, v_0, and y_0 with the appropriate units. **Note:** Since the acceleration is caused by gravity, our notation for it will be a_g rather than just a.

1. The equation of motion with proper units is:

   $y =$

2. The acceleration with proper sign and units is:

   $a_g =$

3. The initial velocity with proper sign and units is:

   $v_0 =$

4. The initial position with proper sign and units is:

   $y_0 =$

Now let's consider the motion of the ball as it falls.

6.3.4. Activity: How a Ball Actually Falls

a. Examine the portion of the graph in Activity 6.3.2b that represents the *downward* motion of the ball. What does the nature of this downward motion look like? Constant velocity, constant acceleration, an increasing or decreasing acceleration? How does your observation compare with the prediction you made in Activity 6.3.1?

b. Using the convention that on the y-axis up is positive and down is negative, is the acceleration of the tossed object positive or negative as it falls (i.e., in what direction is the magnitude of the velocity increasing)?

c. If you think the object is undergoing a constant acceleration, use the modeling technique you used in Unit 4, Activity 4.10.1, to find an equation that describes y as a function of t as the ball falls. **Hints:** You might try to model the system with kinematic equation number 1. Since you are dealing with the second dimension (i.e., the vertical dimension), you should replace x with y and x_0, with y_0. Write the equation of motion in the space below. Then use coefficients of time to find the values of a, v_0 and y_0 with the appropriate units. **Note:** Since the acceleration is caused by gravity, our notation for it will be a_g rather than just a.

1. The equation of motion with proper units is:

$$y =$$

2. The acceleration with proper sign and units is:

$$a_g =$$

3. The initial velocity with proper sign and units is:

$$v_0 =$$

4. The initial position with proper sign and units is:

$$y_0 =$$

Now that you've analyzed the rising and falling motions of the ball, let's put it all together.

6.3.5. Activity: The Acceleration of a Tossed Ball

a. Is the ball's acceleration as it rises the same as or different from its acceleration as it falls? How does this compare to your prediction in Activity 6.3.1? What do you conclude about the acceleration of a tossed ball?

b. Many people are interested in what happens when the ball "turns around" at the top of its trajectory. Some students argue that its acceleration at the top is zero; others think not. What do you think happens to the acceleration at this point?

c. Explain your answer to part b. on the basis of your data, graph, and analysis. **Hint:** Suppose you use the modeling technique on all of your data, instead of separating the data into "rising" and "falling" sections. Does anything special happen at the top of the trajectory?

6.4. WHAT IS GRAVITY?

Hey look, no hands! The object that was tossed experienced an acceleration without the aid of a visible applied force. But if Newton's second law holds, then the net force in the y-direction should equal the mass of the object times its acceleration.

$$\Sigma F_y = ma_g$$

where m is the mass of the object. Maybe a belief in Newton's second law can help us explain the nature of gravity mathematically.

In order to do the next two activities about the nature of gravity you will need the following items:

- 1 balance or electronic scale
- 1 rubber ball
- 1 steel ball
- 1 spring scale, 10 N

Recommended group size:	2	Interactive demo OK?:	N

The Net Force Needed for Your Observed Acceleration

First, you should describe the nature of the force that could cause the acceleration you observed.

6.4.1. Activity: Describing Gravity

a. Use a balance or electronic scale (but not the spring scale) to determine the mass of the object you used in the last set of activities in kg and write it in the space below.

$$m = \_\_\_\_\_\_ \text{ kg}$$

b. Suppose your object was floating in outer space (away from the gravity of the earth, friction, or any other influence) and that Newton's second law holds. Calculate the force in newtons that you would have to apply to your object so it would accelerate as much as the acceleration that you observed in Activity 6.3.2b.

c. If Newton's second law is to be used in the situation where you tossed the object with no visible applied force on it, what force do you need to invent* to make Newton's second law valid? Is the force constant or varying during the time the ball is tossed? What is its magnitude? Its direction? In particular, what is the net force on the object when it is on its way up? At the top of its path? When it is on its way down?

Mass, Acceleration, and Gravitational Influences

So far you have studied the motion of just one object under the influence of the gravitational force you invented or discovered. You should have observed that the acceleration is constant so that the gravitational force is constant. This doesn't tell the whole story. How does the mass of the falling object affect its acceleration? Is the gravitational force constant independent of the mass of the falling object, just the way a horizontal push of your hand might be on a cart that moves in a horizontal direction?

6.4.2. Activity: F_g when Different Masses Fall

a. If you were to drop a massive steel ball and a not very massive rubber ball at the same time, will they fall with the same acceleration? Explain the reasons for your prediction.

b. Release the two objects at the same time. What do you observe? Does it match with your prediction?

c. Use a balance or an electronic scale to determine the mass of the two objects in kg. **Note:** Do not use the spring balance yet.

$$m_{\text{rubber}} = \_\_\_\_\_\_ \text{ kg}$$

$$m_{\text{steel}} = \_\_\_\_\_\_ \text{ kg}$$

* If you already believe Newton's Second Law is an inherent property of nature, then you might prefer to say you discovered the gravitational force. If you feel that you and Newton have been constructing this law on the basis of some interplay between your minds and nature's rules, then you could say you are inventing the idea of the gravitational force.

d. Although you only made a casual qualitative observation of the objects you dropped, it turns out that in the absence of air resistance or other sources of friction all objects accelerate at the rate of $a_g = 9.8$ m/s^2 close to the surface of the earth. There are small variations from place to place and, of course, uncertainties in measurements. If both objects accelerate at the same standard rate, calculate the magnitude of the gravitational force exerted on each one.

$F_{g,\ \text{rubber}} =$ ______ N

$F_{g,\ \text{steel}} =$ ______ N

e. If you have any object of mass m accelerating at a constant rate given by a_g what is the equation that you should use to determine the gravitational force F_g on it? F_g is often referred to as the weight of an object (see below).

f. Check out some weights using a spring scale which has been calibrated in newtons by filling in the chart below.

Object	With balance or electronic scale Mass(kg)	Calculated using Newton's Second Law Fg(N)	Measured by a spring scale Weight (N)
1. Rubber ball			
2. Steel ball			

Fig. 6.5.

6.5. FORCE AND WEIGHT

The Concept of Force–A Review

Force can be defined in several ways which, happily, seem to turn out to be consistent: (1) It can be defined as a push or pull and measured in terms of the stretch of a rubber band or spring; (2) Alternately, the net or combined force on an object can be defined as the cause of motion; (3) Finally, force can be defined in terms of the pull exerted on a mass by the earth as determined by the stretch of a spring when a mass is hanging from it.

Fig. 6.6.

Is There a Difference Between Mass and Weight?
Weight is a measure of the gravitational force, F_g, on a mass m. Mass represents the resistance to motion. Many individuals confuse the concepts of mass and weight. Now that you understand Newton's laws, you should know the difference. Test your understanding by answering the questions posed in the next Activity. Take the fact that gravitational forces on the moon are about 1/6th of those on the Earth.

6.5.1. Activity: Mass and Weight

a. If mass is a measure of the amount of "stuff" in an object, does an astronaut's mass change on the moon? How can astronauts jump so high?

b. Does the astronaut's weight change on the moon? Explain.

c. If weight is a force, what is pushing or pulling? How is weight related to the acceleration of gravity?

d. When did astronauts experience weightlessness? Could they ever experience masslessness?

2D VECTORS AND PROJECTILE MOTION

6.6. EMULATING FALLING MOTIONS BY WHACKING A BALL

Can you learn how to whack a ball so that you can recreate the type of acceleration you have observed for a ball rolling down a ramp or falling freely? If so, we can use the similarity between a falling ball and a whacked ball to study *projectile motion,* in which an object falls vertically and moves horizontally at the same time.

Let's start with one-dimensional measurements. For the measurements described below you will be using a twirling baton with a rubber tip to tap gently on a duck pin ball, which is rather like a small bowling ball. You and a partner should gather the following equipment to study the motion of the ball:

- 1 duck pin ball (or 1 bowling ball)
- 1 twirling baton
- 1 digital stopwatch
- 1 tape measure, 15 m
- 10 bean bags (put dried beans in baby socks)

Recommended group size:	3	Interactive demo OK?:	N

Find a stretch of fairly smooth, level floor that is about 10 meters long. A hallway is a good bet for this series of measurements. You are to record data for position vs. time for your ball for three different situations: (1) a *briskly* rolling ball receiving no whacks, (2) a ball starting at rest and receiving regular light taps, (3) a ball that has an *initial* velocity but is tapped lightly and regularly in the direction opposite to its initial velocity. Before taking data, you and your partner should practice techniques for making these measurements. **Hints:** You should concentrate on making predictions and then taking and recording data for all three types of measurements. Practice coordinating the tapping, timing, and position marking several times before attempting to take data. Data that is not taken carefully will have too much variation to be interpreted easily. *You should obtain at least six values of position.* You may need to spend time after class analyzing the data using spreadsheets and computer graphing.

6.6.1. Activity: The Motion of a Freely Rolling Ball

a. Assume that the ball is rolling freely without friction so there is no net force on it. What do you expect the graphs of position vs. time and of velocity vs. time will look like? Sketch the predicted graphs below. What do you predict the nature of the acceleration of the ball will be? For instance, will it increase, decrease, remain constant, be zero, etc.?

b. Decide how to take and analyze data for position vs. time for a briskly rolling ball that receives no whacks, so as to determine the acceleration of the ball. For the time being, just ignore the friction that causes the ball to stop eventually. Take, analyze, and display your data and findings. Affix a computer printout of your graph in the space below.

c. What is the nature of the acceleration of the ball? Is it increasing, decreasing, constant, zero, etc.? Is that what you predicted? What experimental results are you basing your conclusion on?

6.6.2. Activity: 1D Tapping of a Ball Starting from Rest

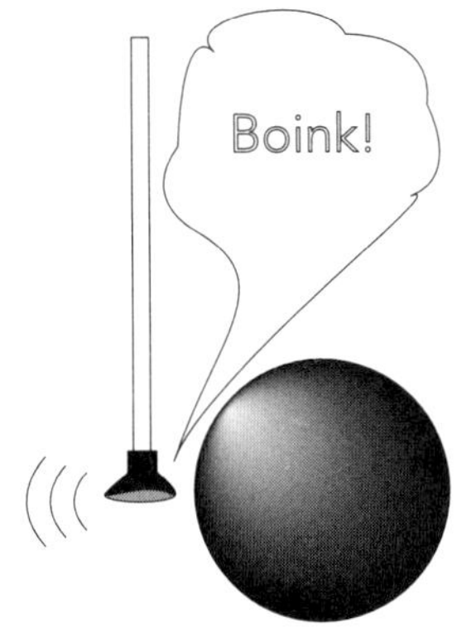

Fig. 6.7.

a. What do you expect the velocity vs. time graph of a ball that starts from rest and receives a series of steady light taps will look like? Sketch the predicted graph below. What do you predict the nature of the acceleration of the ball will be? Will it increase, decrease, remain constant, be zero, etc.?

b. Decide how to take and analyze data for position in meters vs. time in seconds for a ball that is initially at rest and then receives a series of light whacks. Display your data in the following table. Use a spreadsheet to calculate the average velocity of the ball as a function of time, and affix a printout of your spreadsheet in the space below. (You may cover the data table if needed.)

	t(s)	x(m)
1.		
2.		
3.		
4.		
5.		
6.		
7.		
8.		
9.		

	t(s)	x(m)
10.		
11.		
12.		
13.		
14.		
15.		
16.		
17.		
18.		

c. Transfer the spreadsheet data to your graphing routine and create a graph of the average velocity in each time interval as a function of time. Use the slope of the graph to determine the average acceleration of the ball as it experiences continuous tapping. Enclose a printout of the graph in the space below.

d. How did your actual graph of velocity vs. time compare with your predicted graph?

e. What is the nature of the acceleration of the ball? Is it increasing, decreasing, constant, zero, etc.? Is that what you predicted? What experimental results are you basing your conclusion on?

Next, you should set up a situation that is similar to tossing a ball up in the air. To do this, you will need to give the ball an initial velocity in a direction that is opposite to the direction of the force you exert on it.

6.6.3. Activity: 1D Tapping of a Ball with Initial Motion

a. What do you expect the velocity vs. time graph will look like for a ball that is initially rolling toward you as you give it a series of steady light taps in a direction opposite to its initial motion? Sketch the predicted graph below. What do you predict the nature of the acceleration of the ball will be? Will it increase, decrease, remain constant, be zero, etc.? Ignore any slight decreases in velocity that result from small bumps on the floor.

b. Decide how to take and analyze data for position vs. time for a ball that is initially rolling toward you as you give it a series of light whacks that causes it eventually to stop and then turn around. Display your data in the table below. Use a spreadsheet to calculate the average velocity of the ball as a function of time, and affix a printout of it in the space below. (Cover the data table if needed.)

	t(s)	x(m)
1.		
2.		
3.		
4.		
5.		
6.		
7.		
8.		
9.		

	t(s)	x(m)
10.		
11.		
12.		
13.		
14.		
15.		
16.		
17.		
18.		

c. Transfer the spreadsheet data to your graphing routine and create a graph of the average velocity in each time interval as a function of time. Use the slope of the graph to determine the average acceleration of the ball as it experiences continuous tapping. Affix a printout of the graph in the space below.

d. What is the nature of the acceleration of the ball? Is it increasing, decreasing, constant, zero, etc.? Is that what you predicted? What experimental results are you basing your conclusion on? Ignore slight decreases in velocity.

6.7. FITTING EQUATIONS TO YOUR DATA

Now you will further analyze the data you took on the one-dimensional motion of the tapped ball. You should review the modeling techniques described in Unit 4.

If the ball was rolling freely (i.e., experiencing no force) or if you applied a constant force to your ball in the x-direction, the x vs. t graph ought to have a predictable shape.

6.7.1. Activity: Finding Equations to Describe x vs. t

a. Refer to the data you collected in Activity 6.6.1 for the freely rolling ball. Use the modeling techniques you have employed in previous units to fit a line or curve to your data. Affix an overlay graph of your original data and the fitted line or curve in the space below. Also give the equation that best describes how x varies with time.

b. Repeat part a. for the data on the ball tapped from rest that you collected in Activity 6.6.2.

Predicted curve:

Best fit equation:

Printout of graph:

c. Repeat part a. for the data on a ball tapped with initial motion that you collected in Activity 6.6.3.

Predicted curve:

Best fit equation:

Printout of graph:

d. Now look at the equations you obtained in parts a, b, and c. Comment on how the equations are similar and how they differ. Is one of them more different than the others, are all three alike, or are all three quite different? Explain.

6.8. PROJECTILE MOTION–SOME OBSERVATIONS

So far we have been separately dealing with horizontal motion along a straight line and vertical motion along a straight line. In this unit and the next we would like to study motions in a plane such as the motion of a cannon ball or the circular orbit of the planet Venus.

If a rock is hurled off a cliff with an initial velocity in the x-direction and y-direction, it will continue to move forward in the x-direction and at the same time fall in the y-direction as a result of the attraction between the earth and the ball. The two-dimensional motion that results is known as *projectile motion.*

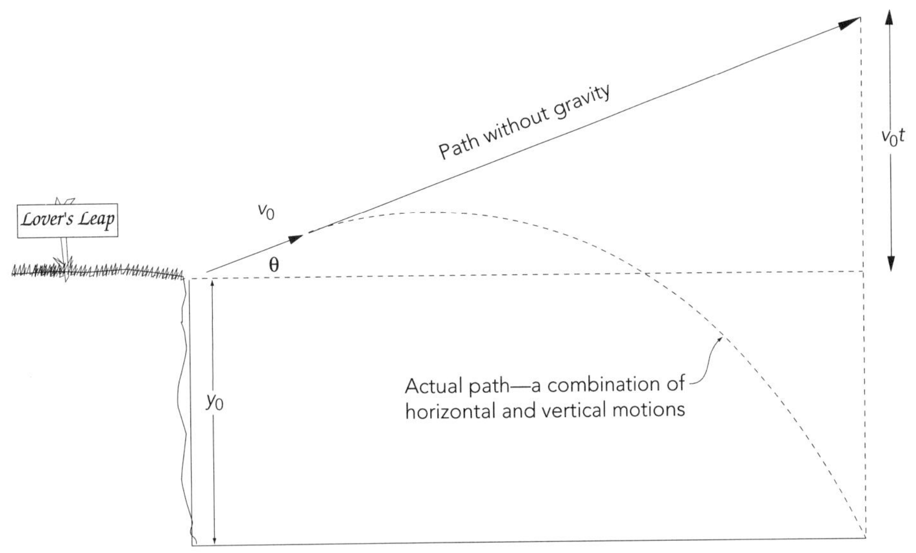

Fig. 6.8. Diagram of a motion with and without gravitational forces present.

In this session you are going to simulate projectile motion by combining the two kinds of one-dimensional motion of the bowling ball that you have already observed. The first motion was that of a ball rolling along with no external forces on it. The second motion was that of a ball receiving a series of rapid light whacks. Once you have set up the motion, you can do a series of quantitative measurements to record the shape of the path described by the ball and measure its position, velocity, and acceleration in the x- and y-directions as the ball progresses. Let's start with some predictions.

6.8.1. Activity: Predicting the Nature of Projectile Motion in a Vertical Plane with F_g Acting

a. Suppose you were to toss a ball at a 45° angle with respect to the horizontal direction. Can you guess what the resulting y vs. x graph of its two-dimensional motion would look like? You can sketch the predicted motion in the graph below. Please use your previous observations of vertical and falling motion in 1D and of horizontal motion in 1D to make an intelligent prediction of the path followed by the ball.

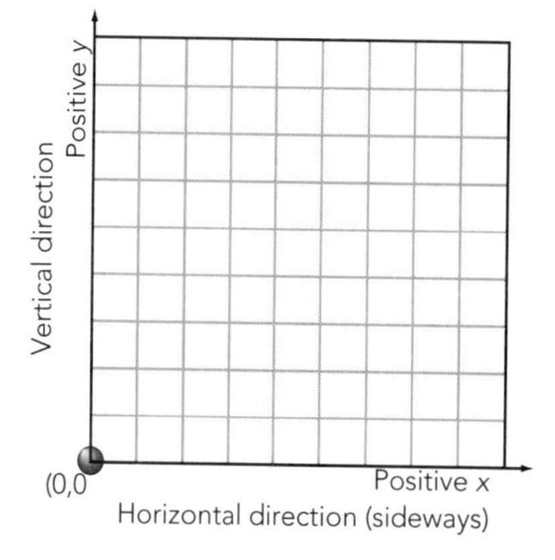

b. Explain the basis for your prediction.

Demonstration of the Independence of Vertical and Horizontal Motion for a Projectile

Suppose someone riding on a moving cart with negligible friction in the wheel bearings tosses a ball straight up in the air. Will the ball fall behind him or will he be able to catch the ball later before it lands?

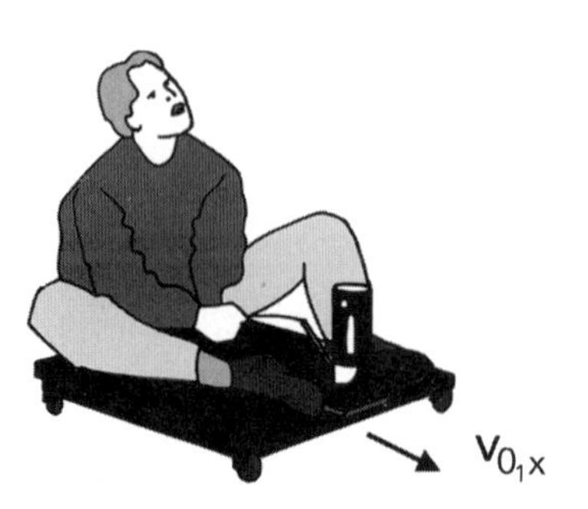

Fig. 6.9. Diagram of a person tossing or launching a ball straight up while moving at a constant velocity horizontally.

For this demonstration you will need:

- 1 Kinesthetic cart
- 1 spring-loaded projectile launcher with a ball

Recommended group size:	All	Interactive demo OK?:	Y

6.8.2. Activity: Predicting the Path of a Ball Tossed from a Moving Cart

a. Describe what you think will happen when a ball is tossed straight up from a moving cart by sketching the path an observer at rest in the laboratory will see.

b. Explain the reasons for your prediction

c. Observe the demonstration and describe what happens.

d. Suppose the cart is moving at 3 m/s. What is the initial horizontal velocity of the ball according to an observer in the laboratory? What is its horizontal velocity a few moments later when it lands on the floor?

e. What happens to the vertical velocity of the ball according to an observer in the laboratory?

f. Do the horizontal and vertical motions seem to be independent?

6.8.3. Activity: Predicting Horizontal Projectile Motion Using an Applied Force

a. Suppose you were to roll the ball briskly in one direction and then proceeded to tap on it at right angles to its original direction. Can you guess what the resulting graph of its two-dimensional motion would look like? You can sketch the predicted motion in the graph below. *Please use your previous observations of vertical and falling motion in 1D and of horizontal motion in 1D to make an intelligent prediction of the path followed by the ball.*

(0,0) Positive *x* axis

Negative *y* axis

b. Explain the basis for your prediction.

6.9. USING VECTORS FOR THE ANALYSIS OF 2D MOTION

Your next task will be to describe projectile motion mathematically. In order to do this, let's take a break from physics and learn about the mathematics of some abstract entities that mathematicians call *vectors.*

The world is full of phenomena that we know directly through our senses—objects moving, pushes and pulls, sights and sounds, winds and waterfalls. A vector is an abstract, mathematical entity that obeys certain predefined rules. It is a mere figment of the mathematician's imagination. But vectors can be used to *describe* aspects of "real" phenomena such as positions, velocities, accelerations, and forces. In the following figure, the position of an object relative to a coordinate system we have chosen can be represented by the vector $\vec{r}$

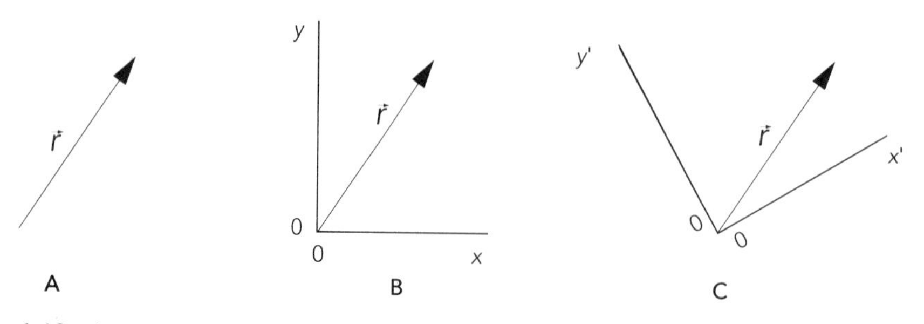

Fig. 6.10. A position vector without a coordinate system and the same vector described by two different coordinate systems with the same origin.

A vector has two key attributes represented by an arrow pointing in space—*magnitude* and *direction*. The *magnitude* of a vector is represented by the length of the arrow and its *direction* is represented by angles between the arrow and the coordinate axes chosen to describe the vector. To answer questions about the magnitude and direction of vector $\vec{r}$ shown in Figure 6.10, you will need:

- 1 ruler
- 1 protractor

Recommended group size:	2	Interactive demo OK?:	N

6.9.1. Activity: Vector Magnitude and Direction

a. What is the magnitude of the position vector in the x, y coordinate system shown in Figure 6.10B?

b. What angle does the vector make with the x-axis in the same coordinate system?

c. What is the magnitude of the vector in the x', y' coordinate system shown in Figure 6.10C? How does this compare to the magnitude of the vector in the x, y coordinate system?

d. What angle does the vector make with the x'-axis? Is this the same as the angle with respect to the x-axis?

Vector Notation

One of the most common ways to indicate that a quantity is a vector is to represent it as a letter with an arrow over it. Thus, in this activity guide symbols such as $\vec{r}$, $\vec{v}$, $\vec{a}$, and $\vec{F}$ all represent quantities which have magnitude *and* direction while r, v, a, and F represent only the magnitudes of those vectors. **Note:** You should *always* place an arrow over vector quantities.

There are some alternate ways to represent vectors. For example, the diagram below illustrates the representation of vectors through unit vector notation.

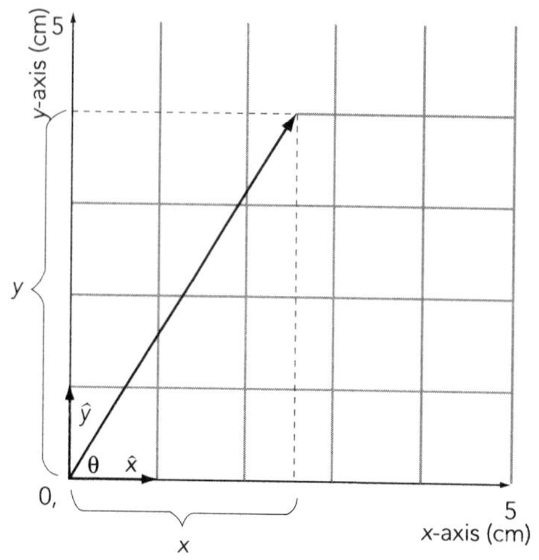

Fig. 6.11. Vector components for a position vector $\vec{r} = x\hat{x} + y\hat{y}$ where $\hat{x}$ and $\hat{y}$ are unit vectors and x and y are vector components.

Reminder: $\hat{x}$ and $\hat{y}$ represent *unit vectors* pointing along the x- and y-axes, respectively. Many texts use $\hat{\imath}$ and $\hat{\jmath}$ and for these quantities.

6.9.2. Activity: Vector Components

a. Use the Pythagorean Theorem to find an equation that relates the magnitude of the vector $\vec{r}$ to the value of its x-component and its y-component, x and y.

b. Using Figure 6.11, measure the value of the vector magnitude, r, and of the vector components, x and y. Don't forget to specify units!

$r_{meas} =$

$x_{meas} =$

$y_{meas} =$

c. Use the measured values of x and y to calculate the magnitude, r, of the vector $\vec{r}$. How does this calculated value compare to the measured value you obtained in part a?

d. Use the definition of sine and cosine in Appendix J to show that if the angle θ is known, then the values of x and y can be calculated from the equations $x = r \cos \theta$ and $y = r \sin \theta$.

e. Measure the value of θ and combine it with the previously measured magnitude of $\vec{r}$ to calculate the values of x and y. How do these compare with the values you measured directly for x and y?

f. Alternatively, show that if the vector components x and y are known, then θ can be determined from the equation $\theta = \arctan (y/x)$.

g. How does the calculated value of θ compare with the measured values?

ANALYZING PROJECTILE MOTION

6.10. DETERMINING THE PATH OF A BALL IN TWO DIMENSIONS

In this activity you will make measurements of "projectile" motion. Then you will use vectors to analyze these measurements.

To make the two-dimensional measurements described below you will be using the twirling baton and duck pin or bowling ball. You and a partner should gather the following equipment to study the motion of the ball:

- 1 duck pin ball (or 1 bowling ball)
- 1 twirling baton
- 1 digital stopwatch
- 1 tape measure, 15 m
- 10 bean bags (put dried beans in baby socks)

Recommended group size:	3	Interactive demo OK?:	N

Find a stretch of fairly smooth level floor that is about 10 meters on a side. A gym floor is a good bet for this series of measurements. You are to record data for position vs. time for a ball that has a brisk initial velocity but is tapped lightly and regularly in the direction perpendicular (i.e., at right angles) to its *initial* velocity. Before taking data, you and your partner should practice techniques for making these measurements. *Try to hit the ball with lots of relatively rapid even taps. Always hit at right angles to the original direction of motion.*

6.10.1. Activity: 2D Tapping of a Big Ball w/ Initial Motion

Record the x and y position of the ball as a function of time and fill in the data table in the space below. *Be sure to describe the procedures you used to take the data.*

No.	t(s)	x(m)	y(m)
1.			
2.			
3.			
4.			
5.			
6.			
7.			
8.			

No.	t(s)	x(m)	y(m)
9.			
10.			
11.			
12.			
13.			
14.			
15.			
16.			

6.11. FITTING EQUATIONS TO YOUR DATA

The main task in this session is to analyze the data you took on the two-dimensional motion of the tapped ball. To do this, you might want to review the section on vectors included in the last session. You should also review the modeling techniques described in Unit 4.

If you applied a constant force to your ball in the y-direction and no force in the x-direction, then the x vs. t and y vs. t graphs ought to have predictable shapes.

6.11.1. Activity: What Equation Describes *x* vs. *t*?

a. Refer to the data you entered in Activity 6.10.1. Transfer your data into a spreadsheet with the following columns in it. Affix it below (you can cover up the list if you like!). *Save your spreadsheet since you will need it in subsequent activities.*

1. Time data
2. Measured x-value
3. Measured y-value (you'll need this in the next activity)

b. Assuming that there is no force in the x-direction, predict the shape of the x vs. t graph and sketch it below.

c. Now graph x vs. t. Does it have the shape you expected?

d. Use the modeling techniques you have used in previous units to fit a line or curve to your data. Affix an overlay graph of your original data and the modeled line or curve in the space below. Also give the equation that best describes how x varies with time.

Note: If there is a (more or less) constant force on the ball in the y-direction, the mathematical relationship between y and t should be a bit more complicated than the one between x and t.

6.11.2. Activity: What Equation Describes *y* vs. *t* ?

a. Return to the spreadsheet you created in Activity 6.11.1. Assuming that there is a constant force in the y-direction, predict the shape of the y vs. t graph and sketch it below.

b. Now graph y vs. t. Does it have the shape you expected?

c. Use the modeling techniques you have used in previous units to fit a line or curve to your y vs. t data. Affix an overlay graph of your original data and the modeled line or curve in the space below. Also give the equation that best describes how y varies with time.

Now, let's combine the equations from Activities 6.11.1d and 6.11.2c to find an equation that describes how y varies with x.

6.11.3. Activity: What Equation Describes y vs. x?

a. Find an equation for y vs. x by combining the equations for x vs. t and y vs. t and eliminating t. Write the equation below.

b. Create a table for actual measured values of y vs. x in your spreadsheet from Activity 6.11.1. Then add a column using your equation from part a. to calculate a *theoretical* value of y for each *measured* value of x. Create an overlay graph showing both measured y vs. x and theoretical y vs. x. Affix your table and graph in the space below.

Name ______________________ Section __________ Date __________

UNIT 7: APPLICATIONS OF NEWTON'S LAWS

This "old-fashioned" record is spinning at $33^1/_3$ revolutions per minute. The coin lying on top of it is spinning round and round with the record. Although it is moving at a constant speed, the coin is actually accelerating. How can that be? If we place the coin farther from the center of the record it will "fly off" in a straight line or if we spin the record faster, say at 45 or 78 revolutions per minute, the penny will also fly off. Why? When you finish this unit you should be able to answer these questions.

UNIT 7: APPLICATIONS OF NEWTON'S LAWS*

The essential fact is that all the pictures which science now draws of nature . . .are mathematical pictures.
Sir James Jeans
(1887–1946)

OBJECTIVES

1. To explore the phenomenon of uniform circular motion and the accelerations and forces needed to maintain it.

2. To use of Newton's laws to describe the effects of gravitational forces in two dimensions.

3. To consider the characteristics of three different types of passive forces: tension (in strings, ropes, springs, and chains), normal forces (which support objects affected by gravity), and friction.

4. To learn to use free-body diagrams to make predictions about the behavior of systems that undergo multiple forces in two and three dimensions.

* Some of the materials and exercises on the use of vector diagrams found in this unit were adapted from Active Learning Problem Sheets (ALPS) developed by Alan Van Heuvelen at New Mexico State University.

7.1. OVERVIEW

In the last unit you began the study of the application of Newton's laws to *projectile motion.* In this unit we are going to consider the application of Newton's laws to several other phenomena in one and two dimensions. Since Newton's laws can be used to predict types of motion or the conditions for no motion, they are used in many endeavors including astrophysics, engineering, and the study of human body motion.

You will begin this unit by exploring *uniform circular motion,* in which an object moves at a constant speed in a circle. In particular, you will develop a mathematical description of centripetal acceleration and the force needed to keep a massive object moving in a circle.

Next you will consider the characteristics of several "invisible" forces that must be taken into account during the comprehensive application of Newton's laws to problems of real interest. These forces—*friction, tension in strings,* and *normal forces*—are called *passive forces* because they only act in response to other forces.

Finally, you will learn techniques for drawing free-body diagrams to help you predict motions or find the conditions for equilibrium (i.e., no motion) in some relatively complicated situations. One of these situations includes the study of motions up and down ramps or inclined planes in which vector components of forces must be considered. In inclined plane motion, the path of the object is constrained to move along a straight line and is not parabolic like that of a projectile.

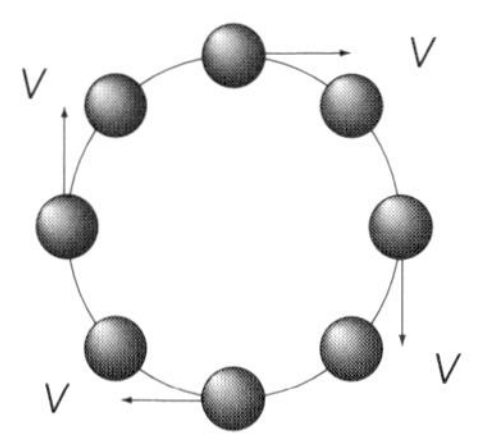

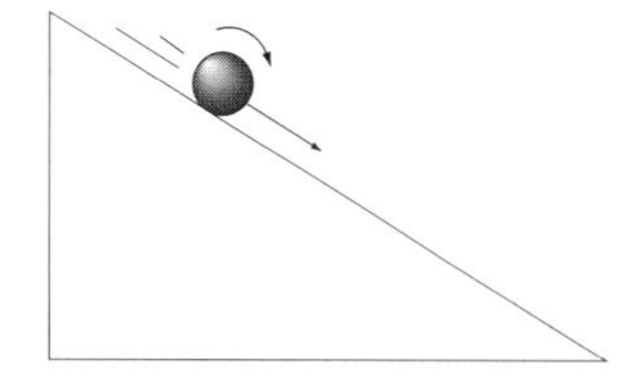

Fig. 7.1. Two types of motion to be explored in this unit: uniform circular motion with constant speed, and frictionless "falling" motion along a straight line that lies along the surface of a ramp.

CIRCULAR MOTION AND CENTRIPETAL FORCE

7.2. MOVING IN A CIRCLE AT A CONSTANT SPEED

When a race car speeds around a circular track, or when David twirled a stone at the end of a rope to clobber Goliath, or when a planet like Venus orbits the sun, they undergo *uniform circular motion.* Understanding the forces that govern orbital motion has been vital to astronomers in their quest to understand the laws of gravitation.

But we are getting ahead of ourselves, for as we have done in the case of linear and projectile motion we will begin our study by considering situations involving external applied forces that lead to circular motion in the absence of friction. We will then use our belief in Newton's laws to see how the circular motions of the planets can be used to help astronomers discover the laws of gravitation.

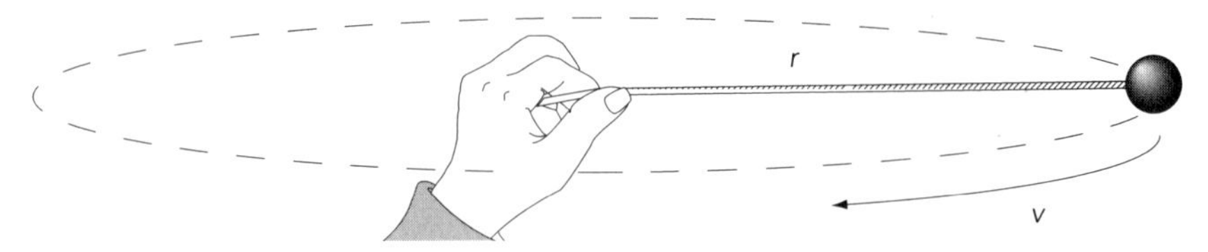

Fig. 7.2. Uniform circular motion. A ball moving at a constant speed in a circle of radius *r*.

Let's begin our study with some very simple considerations. Suppose an astronaut goes into outer space, ties a ball to the end of a rope, and spins the ball so that it moves at a constant speed.

7.2.1. Activity: Uniform Circular Motion

a. Consider Figure 7.2. What is the speed of a ball that moves in a circle of radius $r = 2.5$ m if it takes 0.50 s to complete one revolution?

b. The *speed* of the ball is constant! Would you say that this is accelerated motion? Why or why not?

c. What is the *definition* of acceleration? (Remember that acceleration is a vector!)

d. Are *velocity* and *speed* the same thing? Is the velocity of the ball constant? (**Hint:** Velocity is a vector quantity!)

e. In light of your answers to c. and d., would you like to change your answer to part b? Explain.

7.3. USING VECTORS TO DIAGRAM HOW VELOCITY CHANGES

By now you should have concluded that since the *direction* of the motion of the ball is constantly changing, its velocity is also changing and thus it is accelerating. We would like you to figure out how to calculate the *direction* of the acceleration and its magnitude as a function of the speed v of the ball as it revolves and as a function of the radius of the circle in which it revolves. In order to use vectors to find the direction of velocity change in circular motion, let's review some rules for adding velocity vectors.

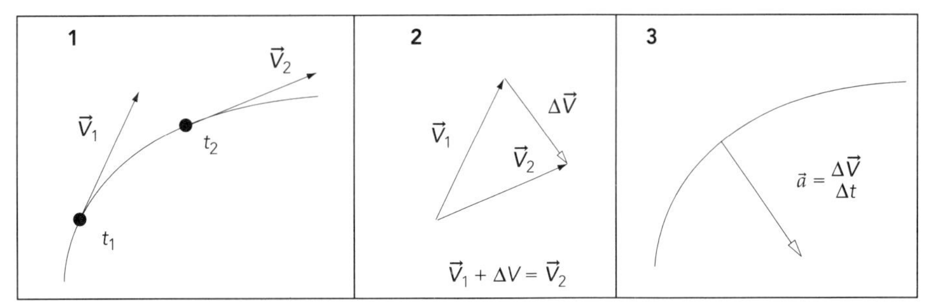

Fig. 7.3. 1. *To draw a velocity vector:* Draw an arrow representing the velocity, $\vec{v}_1$, of the object at time t_1. Draw another arrow representing the velocity, $\vec{v}_2$, of the object at time t_2.

2. *To represent velocity change:* Find the change in the velocity during the time interval $\Delta t = t_2 - t_1$. You can start by using the rules of vector sums to rearrange the terms so that $\Delta\vec{v} + \vec{v}_1 = \vec{v}_2$. To draw the vectors involved, place the tails of the two velocity vectors together halfway between the original and final location of the object. The change in velocity is the vector that points from the head of the first velocity vector to the head of the second velocity vector.

3. *To represent acceleration:* The acceleration equals the velocity change $\Delta\vec{v}$ divided by the time interval Δt needed for the change. Thus, it is in the same direction as Δv but is a different length unless ($\Delta t = 1$). Thus, even if you do not know the time interval, you can still determine the direction of the acceleration because it points in the same direction as $\Delta\vec{v}$.

The acceleration associated with uniform circular motion is known as *centripetal acceleration.* You should use the vector diagram technique to find its direction.

7.3.1. Activity: Centripetal Acceleration Direction

a. Determine the direction of motion of the ball shown below if it is moving counterclockwise at a constant speed. Note that the direction of the ball's velocity is always tangential to the circle as it moves around. Draw an arrow representing the direction and magnitude of the ball's velocity as it passes the dot *just before* it reaches point A. Label this vector $\vec{v}_1$.

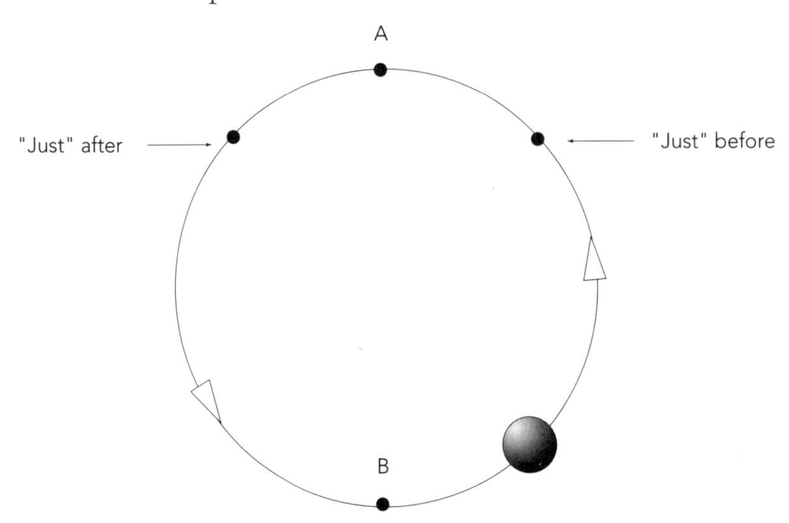

b. Next, use the same diagram to draw the arrow representing the velocity of the ball when it is at the dot just after it passes point A. Label this vector $\vec{v}_2$.

c. Find the direction and magnitude of the change in velocity as follows. In the space below, make an exact copy of both vectors, placing the tails of the two vectors together. (See Figure 7.3, diagram 2.) Next, draw the vector that must be added to vector $\vec{v}_1$ to add up to vector $\vec{v}_2$; label this vector $\Delta\vec{v}$. *Be sure that vectors $\vec{v}_1$ and $\vec{v}_2$ have the same magnitude and direction in this drawing that they had in your drawing in part (a)!* (Again, see Figure 7.3.)

d. Now, draw an exact copy of $\Delta\vec{v}$ on your sketch in part a. Place the tail of this copy at point A. Again, make sure that your copy has the exact magnitude and direction as the original $\Delta\vec{v}$ in part c.

e. Now that you know the direction of the change in velocity, what is the direction of the centripetal acceleration, $\vec{a}_c$?

f. If you redid the analysis for point B at the opposite end of the circle, what do you think the direction of the centripetal acceleration, $\vec{a}_c$, would be now?

g. As the ball moves on around the circle, what is the direction of its acceleration?

h. Use Newton's second law in vector form ($\Sigma\vec{F} = m\vec{a}$) to describe the direction of the net force on the ball as it moves around the circle.

i. If the ball is being twirled around on a string, what is the source of the net force needed to keep it moving in a circle?

7.4. USING MATHEMATICS TO DERIVE HOW CENTRIPETAL ACCELERATION DEPENDS ON RADIUS AND SPEED

You haven't done any experiments yet to see how centripetal acceleration depends on the radius of the circle and the speed of the object. You can use an understanding of Newton's second law to get a feel for what the mathematical relationships might be. You can then use the rules of mathematics and the definition of acceleration to *derive* the relationship between speed, radius, and magnitude of centripetal acceleration.

7.4.1. Activity: How Does a_c Depend on v and r?

a. Do you expect you would need more centripetal acceleration or less to cause an object moving at a certain speed to rotate in a smaller circle? In other words, would the magnitude, a_c, have to increase or decrease as r decreases if circular motion is to be maintained? Explain.

b. Do you expect you would need more centripetal acceleration or less to cause an object to rotate at a given radius r if the speed v is increased? In other words, would the magnitude, a_c, have to increase or decrease as v increases for circular motion to be maintained? Explain.

You should have guessed that it requires more acceleration to move an object of a certain speed in a circle of smaller radius and that it also takes more acceleration to move an object that has a higher speed in a circle of a given radius. Let's use the definition of acceleration in two dimensions and some accepted mathematical relationships to show that the magnitude of centripetal acceleration, a_c, should actually be given by the equation

$$a_c = \frac{v^2}{r} \tag{7.1}$$

To do this derivation you will want to use the following definition for acceleration

$$\langle \vec{a} \rangle \equiv \frac{\vec{v}_2 - \vec{v}_1}{t_2 - t_1} = \frac{\Delta \vec{v}}{\Delta t} \tag{7.2}$$

7.4.2. Activity: Finding the Equation for a_c

a. Refer to the diagram that follows. Explain why, at the two points shown on the circle, the angle between the position vectors at times t_1 and t_2 is the same as the angle between the velocity vectors at times t_1 and t_2. **Hint:** In circular motion, velocity vectors are always perpendicular to their position vectors.

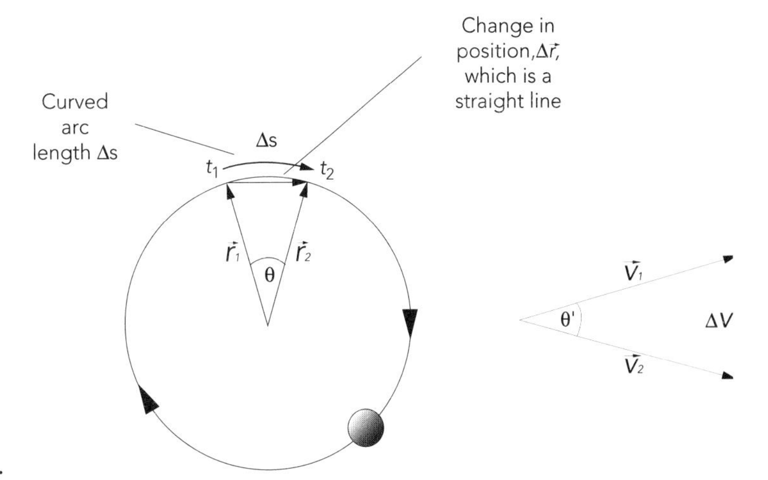

Fig. 7.4.

b. Since the angles are the same and since the magnitudes of the positions never change (i.e., $r = r_1 = r_2$) and the magnitudes of the velocities never change (i.e., $v = v_1 = v_2$), use the properties of similar triangles to explain why

$$\frac{\Delta v}{v} = \frac{\Delta r}{r}$$

c. Now use the equation in part b and the definition of $\langle a \rangle$ to show that $\langle a_c \rangle = \frac{\Delta v}{\Delta t} = \frac{(\Delta r)v}{\Delta t r}$.

d. The speed of the object as it rotates around the circle is given by

$$v = \frac{\Delta s}{\Delta t}$$

Is the change in *arc length*, Δs, larger or smaller than the magnitude of the change in the *position* vectors, Δr? Explain why the arc length change and the change in the position vector are approximately the same when Δt is very small (so that the angle θ becomes very small)—that is, why is $\Delta s \approx \Delta r$?

e. If $\Delta s \approx \Delta r$, what is the equation for speed in terms of Δr and Δt?

f. Using the equation in part c, show that as $\Delta t \to 0$, the instantaneous value of the centripetal acceleration is given by

$$a_c = \frac{v^2}{r}$$

g. If the object has a mass m, what is the equation for the magnitude of the centripetal force needed to keep the object rotating in a circle (in terms of v, r, and m)? In what direction does this force point as the object rotates in its circular orbit?

7.5. EXPERIMENTAL VERIFICATION OF THE CENTRIPETAL FORCE EQUATION

The theoretical considerations in the last activity should have led you to the conclusion that, whenever you see an object of mass m moving in a circle of radius r at a constant speed v, it must at all times be experiencing a net centripetal force directed toward the center of the circle, which has a magnitude of

$$F_c = ma_c = m\frac{v^2}{r} \tag{7.3}$$

Let's check this out. Does this rather odd equation really work for an external force? To do this experiment you will need the following equipment:

- 1 smooth, level area (about 2 m in radius)
- 1 post in the center of the area
- 1 2D cart (with bearings to support 2D motion)
- 1 belt with rings on the sides for the rider
- 3 ropes, 3 m (smooth nylon ropes are recommended)
- 1 spring scale, 150 N
- 1 stopwatch
- 1 bathroom scale (for determining cart and rider mass)
- 1 rod, 1.5" dia. × 3' (for finding rider balance point)

Recommended group size:	3	Interactive demo OK?:	Y

You should attach the spring scale between the center post and the rope. The end of the rope should be attached to the cart rider's belt or held by the rider. The second rope should be attached to the cart in a direction that is perpendicular to the original rope. Use the aluminum rod to determine the distance from the center of the circle to the balance point of the rider. Three people are needed for this experiment—a rider, a puller, and a scale reader. See the following diagram.

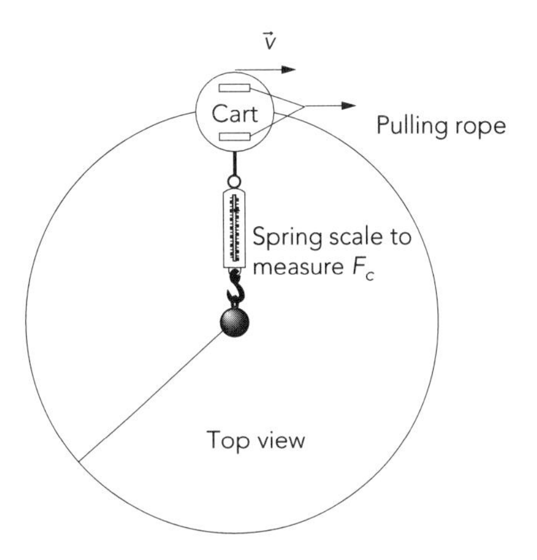

Fig. 7.5. Top view of centripetal force experiment.

7.5.1. Activity: Verifying the F_c Equation

a. If a puller applies a force on the cart that is always tangent to the circle and just sufficient to overcome friction in the cart and maintain the cart's motion at a constant speed, what is the *net* force in a direction perpendicular to the circle? Remember Newton's first law!!??

b. You and members of your group should practice pulling each of your members around at a constant speed. The rider doesn't need the belt for this exercise. Instead, the rider should hold onto the rope and close his or her eyes and concentrate on feeling the centripetal force.

c. When you hold onto the center rope while riding on the cart, what is the direction of the *net* force you feel on you (include consideration of the forces in the radial direction and in the tangential direction)? Does the force you feel seem to increase as you rotate faster? As you rotate in a smaller circle? Explain.

d. Next, keep the radius to the center of the cart and the mass of the cart and rider constant and take data for several different orbital speeds with the rope attached to the belt that the rider is wearing. Set up a spreadsheet and compare the theoretical value of F_c calculated using equation 7.3 with the experimental value obtained by reading the average force in newtons recorded on the spring scale. Use the space below to explain your procedures, show sample calculations, and present your results. **Note:** When reading the force scale you should record the amount of wobble around an average value and use it as an estimate of the most significant uncertainty in your measurements.

Basic data	
Mass of cart (lb)	
Mass of rider (lb)	
Total mass rotating (kg)	
Radius of circular path (m)	
Circumference of path (m)	

	Time/rev t(s)	Average speed (m/s)	Spring scale reading (kg)	Meas. avg. F_c (N)	≈ Uncertainty $\pm\Delta F_c$ (N)	Calculated F_c(N)
1.	∞	0.0	0.0	0.0	0.0	0.0
2.						
3.						
4.						
5.						

e. Plot a graph of F vs. v using a polynomial fit. Affix the graph in the space below. **Hint:** According to Equation 7.3, what order should the polynomial relationship be?

f. Draw vertical error bars over each of your data points to indicate the estimated uncertainty in your force measurements. (See Appendix D for information about what error bars look like.) Then look at the visual data fit. Within the limit of experimental uncertainty, how well does your experimental data support the hypothesis that F_c is a function of v_2?

g. Let's look at agreement between experiment and theory another way. How well does each measured force agree with the corresponding calculated force shown in your data table?

h. Discuss the major sources of uncertainty in this experiment. There are plenty!

7.6. ESSAY: NEWTON'S APPLE

Newton used the idea of centripetal acceleration and the observation that the earth moves in an approximately circular orbit around the sun to conclude that the earth's orbit is a result of a *gravitational force* between the sun and the earth.

Richard Westfall, one of Newton's biographers, asserted that even if Newton had died in 1684, before completing the Principia, the world would have honored him as a genius for his work in mechanics and optics.* However, most intellectual historians consider Newton's most profound contribution to science to be his hypothesis published in the Principia that the same laws of mechanics govern the orbital motion of planets, the falling of objects near the earth, and the linear acceleration of objects that are pushed or pulled with constant forces.

Fig. 7.6. What Newton might have said.
...there is clearly a dynamic identity between uniform circular motion and constantly accelerated motion...

* Richard S. Westfall, *Never at Rest: A Biography of Issac Newton* (Cambridge University Press, Cambridge, 1980).

The image of Newton having this idea as a sudden insight while sitting under an apple tree is probably mythical. However, it leads to some interesting modern facts. Nabisco now distributes both Fig and Apple Newtons. Near the surface of the earth the force on a small apple is about one Newton. Think of this the next time you eat Apple Newtons!

NEWTON'S THIRD LAW AND PASSIVE FORCES

7.7. AN INTRODUCTION TO NEWTON'S THIRD LAW

In order to apply Newton's laws to complex situations with strings, pulleys, inclined planes, and so forth, we need to consider a third force law formulated by Newton having to do with the forces of interaction between two objects. To "discover" some simple aspects of the third law, you should make some straightforward observations using the following equipment:

- 2 spring scales
- 1 Kinesthetic cart (human-sized)

Recommended group size:	2	Interactive demo OK?:	Y

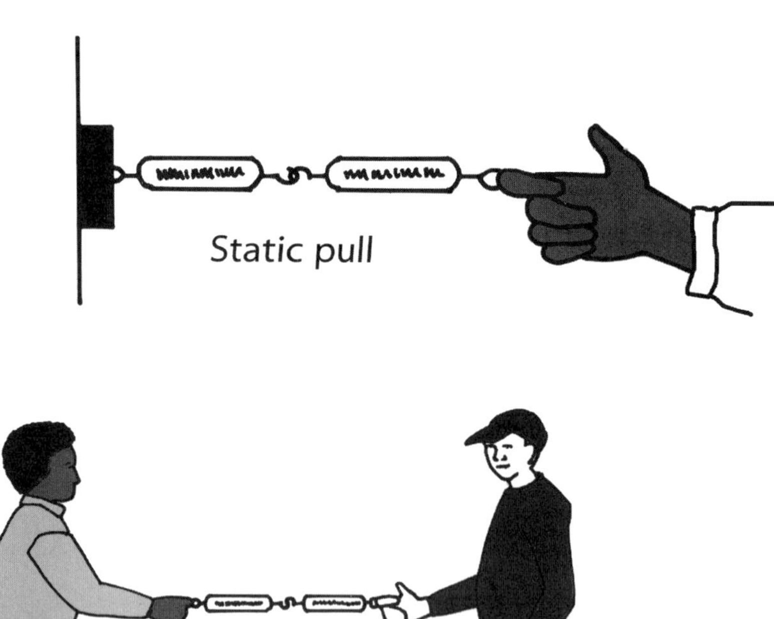

Fig. 7.7. Ways to test Newton's third law.

7.7.1. Activity: Newton's Third Law–Forces of Interaction

Set up the situations shown in the diagram above and see if there are any circumstances in which the object that is pulling and the object that is being pulled exert *different* forces on each other. Describe your conclusions below. **Note:** You can use a skater, a small cart, a person riding on a large cart, etc. for your dynamic observations.

Newton's Third Law

In contemporary English, Newton's third law can be stated as follows:
If one object exerts a force on a second object, then the second object exerts a force back on the first object that is equal in magnitude and opposite in direction to that exerted on it by the first object.

In mathematical terms, using vector notation, we would say that the forces of interaction of object 1 on object 2 are related to the forces of interaction of object 2 on object 1 as follows:

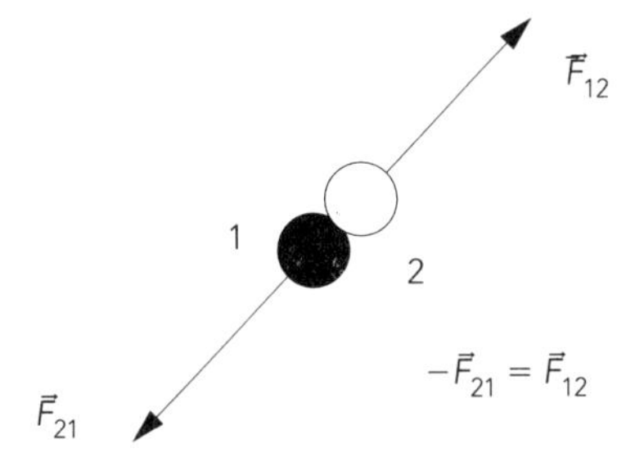

Fig. 7.8. "Equal" and "opposite" force vectors signifying forces that are equal in magnitude and opposite in direction.

Newton actually formulated the third law by studying the interactions between objects when they collide. It is difficult to understand the significance of this law fully without first studying collisions. We will consider this law again in the study of collision processes.

7.8. TENSION FORCES

When you pull on one end of a rope attached to a crate, a force is transmitted down the rope to the crate. If you pull hard enough, the crate may begin to slide. *Tension* is the name given to forces transmitted in this way along devices that can stretch such as strings, ropes, rubber bands, springs, and wires.

Fig. 7.9. Transmitting forces along a string by pulling.

The end of a rope tied to the crate can apply a force to the crate *only* if you first pull on the other end of the rope. Thus, both tension and friction are *passive forces* that only act in reaction to active forces. The characteristics of tension are very different than those of friction. In order to analyze situations in which objects are attached by strings, rubber bands, or ropes it is necessary to understand some attributes of tension forces. You should be able to answer the following related questions:

1. What is the mechanism for creating tension in strings, ropes, and rubber bands?
2. If a string exerts a tension force on an object at one end, what is the magnitude and direction of the tension force it exerts on another object at its other end?
3. What happens to the magnitude and directions of the tension forces at each end of a string and in the middle of that string when the direction of the string is changed by a post or pulley?
4. Can a flexible force transmitter (like a string) support a lateral (or sideways) force?

In order to investigate the nature of tension forces you will need the following apparatus:

- 4 #16 rubber bands
- 4 strings, approximately 25 cm each (with small loops at the ends)
- 4 strings, approximately 50 cm each (with small loops at the ends)
- 3 spring scales, 5 N (matched)
- 3 spring scales, 20 N (matched)
- 2 spring scales, 200 N (matched)
- 2 ropes, approximately 2 m each
- A set of small hanging masses
    - 2 masses, 100 g
    - 2 masses, 200 g

  - 2 masses, 1000 g
  - 1 mass, 10 kg
- 2 table clamps
- 2 low friction pulleys
- 2 right angle clamps
- 2 rods

Recommended group size:	3	Interactive demo OK?:	Y

Mechanisms for Tension and the Direction of Forces

For these observations you should stretch a rubber band and then a string between your hands as shown in the diagram below. First, just feel the directions of the forces. Then add the spring scales and both feel and measure the forces.

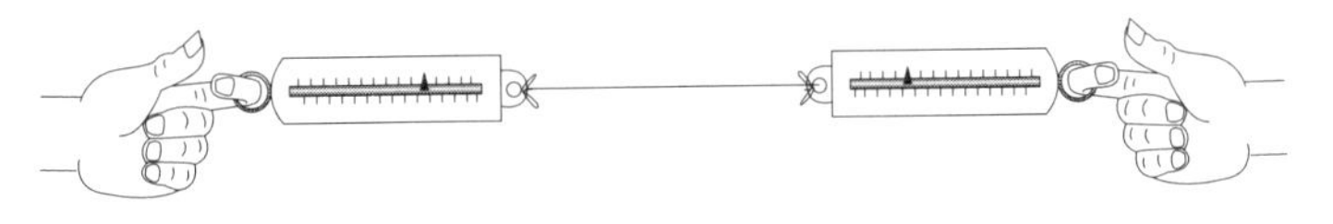

Fig. 7.10. Stretching a string.

7.8.1. Activity: Tension Mechanisms and Force Directions

a. Pull on the two ends of a rubber band. (Forget about the spring scales for now). Does the rubber band stretch? What is the direction of the force applied by the rubber band on your right hand? On your left hand?

b. Does the magnitude of the forces applied by the rubber band on each hand feel the same?

c. Repeat this activity with a string instead of a rubber band. This time, use a spring scale at each end to measure the forces at the ends of the string. Does the string stretch? (Look carefully!)

d. If you pull by the same amount on the string as you did on the rubber band, does substituting the string for the rubber band change anything about the directions and magnitudes of the tension forces exerted on each hand?

e. If the forces caused by the string on your left and right hands respectively are given by $\vec{F}_{T1}$ and $\vec{F}_{T2}$, what is the equation that relates these two forces?

f. What is a mechanism that might cause a rubber band or a string to develop tension in response to a force that you apply?

Tension and Newton's Third Law

Suppose you find yourself in the situations depicted that follow. In which situation would you feel the most discomfort, A or B? After you make a prediction, you can work with two other partners and use two ropes and two large spring scales (200 N or greater) to check your predictions.

Fig. 7.11a. Being pulled in two directions by strong men.

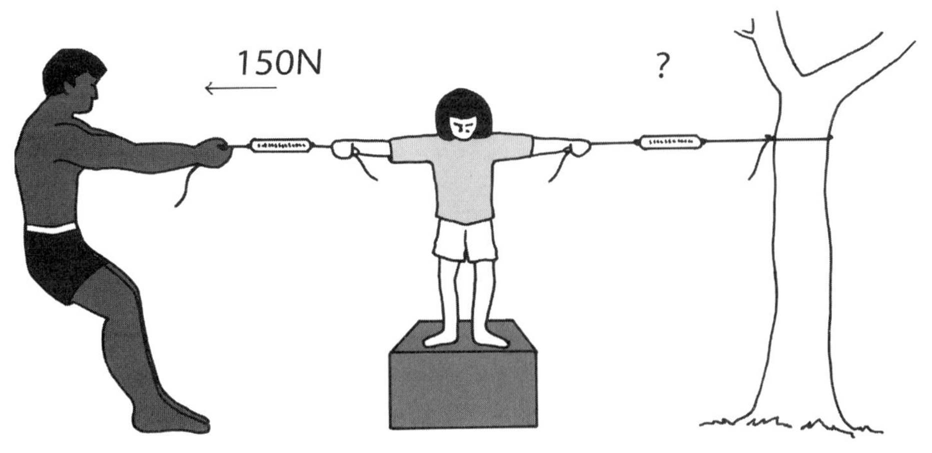

Fig. 7.11b. Being pulled in two directions by a strong man and a tree.

7.8.2. Activity: Tension and Newton's Third Law

a. Which situation do you expect to cause you the most discomfort, a. or b? Explain the reasons for your prediction.

b. Talk to your partners or have a discussion with members of your class. Do you want to change your prediction? If so, explain the reasons for your new prediction.

c. Work with two other reasonably strong partners. Use large spring scales and ropes to test your predictions. Describe what happens. Which situation is most uncomfortable?

Based on Newton's third law and the observations you just made, answer the following questions using vector notation and arrows. Assume the strong man in the diagram below is pulling to the left on a rope with a force of magnitude 150 N.

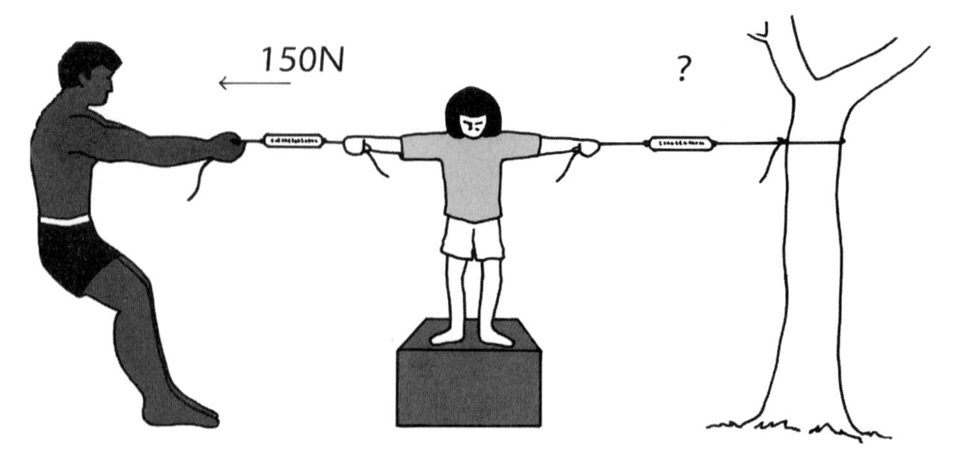

Fig. 7.12. Being pulled in two directions by a strong man and a tree.

7.8.3. Activity: Figuring out the Rope Tensions

a. What is the magnitude and direction of the force that the rope is exerting on the man? Draw this force vector both in the space below and on Figure 7.12 and mark it with an "a."

b. What force is the left-hand rope exerting on the girl's left arm? Draw this force vector both in the space below and on Figure 7.12 and mark it with a "b."

c. What force is the spring scale experiencing on its left end? Draw this force vector both in the space below and on Figure 7.12 and mark it with a "c."

d. What force is the spring scale experiencing on its right end? Draw this force vector both in the space below and on Figure 7.12 and mark it with a "d."

e. What is the reading on the spring scale?

f. What force is the rope exerting on the tree? Draw this force vector both in the space below and on Figure 7.12 and mark it with an "f."

g. What force is the tree exerting on the rope? Draw this force vector both in the space below and on Figure 7.12 and mark it with a "g."

Tension Forces When a String Changes Direction

Suppose you were to hang equal masses of m = 1.0 kg in the various configurations shown below. Predict and measure the tension in the string for each of the situations.

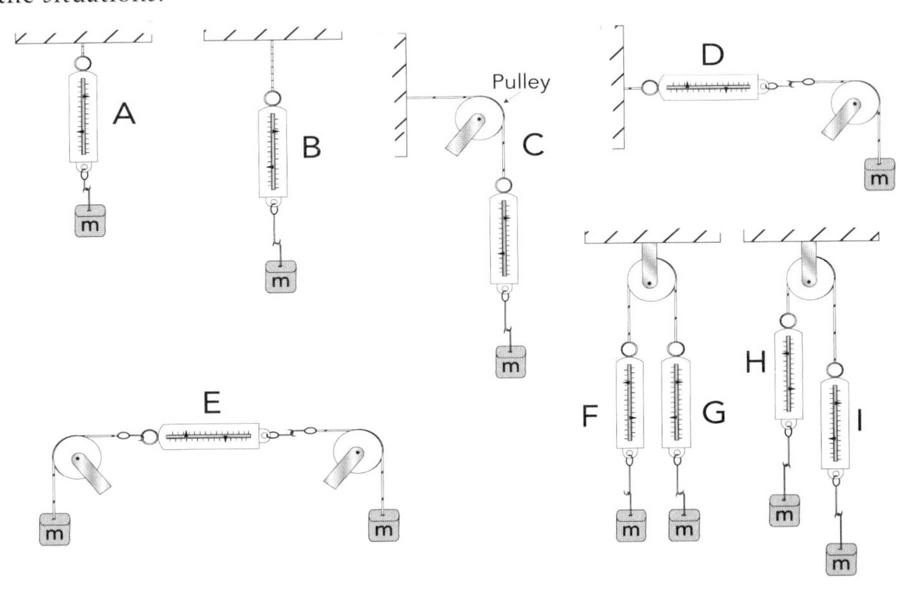

Fig. 7.13. Configurations for measuring tension in a string.

7.8.4. Activity: Tension and Direction Changes

a. For each configuration in Figure 7.14, predict the reading in newtons on each of the spring scales; these readings indicate the forces transmitted by the tensions at various places along the string. Then measure all the forces and record their values. Remember that $m = 1.0$ kg.

Predicted force magnitudes	Measured force magnitudes
$F_A =$ ______ N	$F_A =$ ______ N
$F_B =$ ______ N	$F_B =$ ______ N
$F_C =$ ______ N	$F_C =$ ______ N
$F_D =$ ______ N	$F_D =$ ______ N
$F_E =$ ______ N	$F_E =$ ______ N
$F_F =$ ______ N	$F_F =$ ______ N
$F_G =$ ______ N	$F_G =$ ______ N
$F_H =$ ______ N	$F_H =$ ______ N
$F_I =$ ______ N	$F_I =$ ______ N

b. Summarize what your observations reveal about the nature of tension forces everywhere along a string.

Can a String Support Lateral Forces?

Take a look at the diagram below. Can the strongest member of your group stretch a string or rope so that it is perfectly horizontal when a 10-kg mass is hanging from it? In other words, can the string provide a force that just balances the force exerted by the mass?

Fig. 7.14. Can you raise a 10-kg mass so the ropes are horizontal and the angle θ is zero?

7.8.5. Activity: Can a String Support a Lateral Force?

a. Draw a vector diagram showing the directions of the forces exerted by the strings on the mass hook in the diagram below. What would happen to the direction of the forces as θ goes to zero? Do you think it will be possible to support the mass when $\theta = 0$?

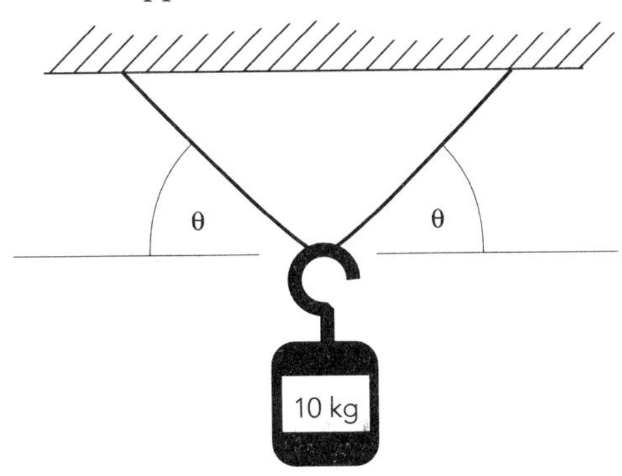

b. Now, experiment with holding a mass horizontally with a string. What do you conclude about the ability of a string to support a mass having a force that is perpendicular to the direction of the string?

7.9. USING TENSION CONCEPTS WITH AN ATWOOD'S MACHINE

Sometime before 1780, a physicist at Cambridge University named George Atwood devised a marvelous machine for measuring the acceleration of a falling mass without the aid of high-speed timers, motion detectors, or video cameras. It consists of two masses connected to each other by means of a light string passing over a relatively frictionless light pulley, as shown in Fig. 7.15.*

* In referring to a "light" pulley and string, we mean that the mass of these items is very small compared to the masses of the falling weights. Thus, the masses of the string and pulley can be neglected in any calculations. Physicists are often kidded about their massless strings and pulleys!

Atwood's machine is not only historically important, it also allows us to practice applying Newton's laws and the kinematic equations to the analysis of motion. In order to make observations of the motion of the masses on an Atwood's machine you will need:

- 1 low-friction and low-mass pulley
- 1 string
- 2 mass holders, 50 g
- 4 table clamps
- 3 rods (to support the pulley)
- 2 masses, 20 g

Recommended group size:	2	Interactive demo OK?:	N

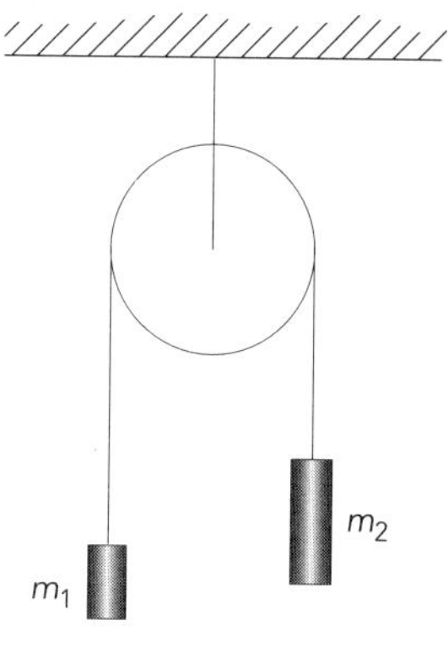

Fig. 7.15. An Atwood's machine with $m_2 > m_1$.

7.9.1. Activity: Behavior of the Atwood's Machine

a. Assume that the two masses are equal. If you pull down gently on one of them, what motion do you predict will result? Explain your reasoning.

b. Set up the Atwood's machine with combinations of equal masses, pull on one of them gently, and describe what you observe. How does your observation compare with your prediction?

c. Suppose that m_2 is greater than m_1. What do you expect to observe and why?

d. Set up the Atwood's machine with combinations of unequal masses and describe what you observe. How do your observations compare with your prediction?

e. If the tension on the string is denoted by F_T, draw a diagram describing the forces on m_1. Draw another diagram showing the forces on m_2. **Hint:** Include the gravitational forces acting in the negative y-direction on masses 1 and 2 and the tension forces acting upwards. Assume that m_2 is greater than m_1.

f. Write down an equation for the net force on m_1 in terms of F_T, g, and m_1. Use Newton's second law to relate this net force to the acceleration, a_1, of m_1.

g. Next, write down an equation for the net force on m_2 in terms of F_T, a_g, and m_2. Use Newton's second law to relate this net force to the acceleration a_2, of m_2.

h. Why can we say the acceleration of m_2 is $a_2 = -a$ if the acceleration of m_1 is $a_1 = a$?

i. Eliminate F_T from the two equations to show that the net acceleration for the Atwood's machine is given by the equation

$$a = \frac{m_2 - m_1}{m_2 + m_1} a_g$$

You have just derived the famous Atwood's equation.

This equation may seem pretty arcane at first glance, but it could end up saving your life if you're a bricklayer! Bricklayers have been known to use pulleys to haul bricks to the upper floors of buildings. This led a British humorist, Gerald Hoffnüng, to compose the following song.

The Bricklayer's Song

Fig. 7.16.

Dear sir, I write this note to you to tell you of my plight,
For at the time of writing it I'm not a pretty sight;
My body is all black and blue, my face a deathly gray,
And I write this note to say why I am not at work today.

While working on the 14th floor, some bricks I had to clear,
But tossing them down from such a height was not a good idea.
The foreman wasn't very pleased; he is a rigid hack
And he said I had to cart them down the ladder on my back.

Well, clearing all these bricks by hand – it was so very slow,
So I hoisted up a barrel and secured a rope below.
But in me haste to do the job, I was too blind to see
That a barrel full of building bricks was heavier than me.

And so when I untied the rope the barrel fell like lead,
And clinging tightly to the rope, I started up instead.
I shot up like a rocket and to my dismay I found
That halfway up I met the bloody barrel coming down.

Well, the barrel broke me shoulder as to the ground it sped,
And when I reached the top I banged the pulley with me head.
But I clung on tightly, numb with shock, from this almighty blow,
While the barrel spilled out half its bricks, some 14 floors below.

Now when these bricks had fallen from the barrel to the floor
I then outweighed the barrel, and so started down once more.
But I clung on tightly to the rope, me body racked with pain
And half way down I met the bloody barrel once again.

The force of this collision half way down the office block
Caused multiple abrasions and a nasty case of shock,
But I clung on tightly to the rope as I fell towards the ground,
And I landed on the broken bricks the barrel scattered 'round.

Well as I lay there on the floor I thought I'd passed the worst,
But the barrel hit the pulley wheel and then the bottom burst.
A shower of bricks rained down on me; I didn't have a hope,
As I lay there bleeding on the ground I let go the bloody rope.

The barrel now being heavier, it started down once more.
It landed right across me as I lay there on the floor.
It broke three ribs and my left arm, and I can only say
I hope you'll understand why I am not at work today.

Now the moral of my story it is awfully plain to see
That physics is a class that would have been some help to me;
So study all your lessons well and think before you act,
Or you'll run the risk of suffering from your own mistakes' impact.

by N. Murphy

7.10. NORMAL FORCES

A book resting on a table does not move; neither does a person pushing against a wall. According to Newton's first law, the net force on the book and on the person's hand must be zero. We have to invent another type of passive force to explain why books don't fall through tables and hands don't usually punch through walls. The force exerted by any surface always seems to act in a direction perpendicular to that surface; such a force is known as a *normal force*. Normal forces are passive forces because they seem to act in response to active forces like pushing forces and gravitational forces. To investigate normal forces you will need the following apparatus:

- 2 meter sticks, 1 m each
- 1 wall
- 1 mass, 50 g
- 1 mass, 100 g
- 2 textbooks, >3" thick

Recommended group size:	3	Interactive demo OK?:	N

By applying forces perpendicular to flexible surfaces with different degrees of stiffness, you can discover a mechanism for the passive normal forces that crop up in reaction to active forces.

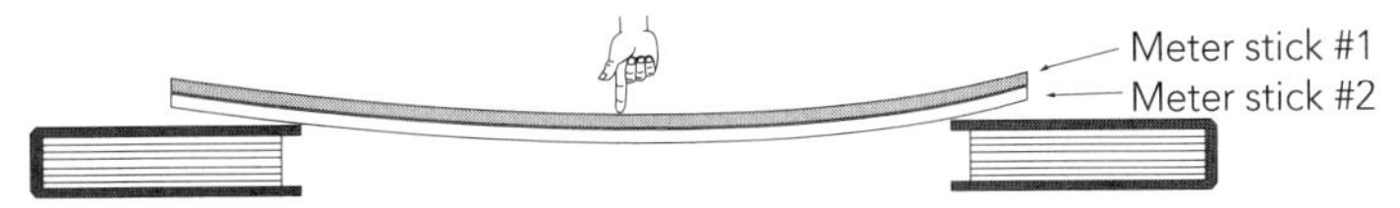

Fig. 7.17. Pressing down on two meter sticks suspended between textbooks.

7.10.1. Activity: Normal Forces

a. Stack one meter stick on top of another between two books to create a "flexible" surface. Press the center of the meter sticks with your finger perpendicular to the surface of the top one. What does your finger feel? What happens to the meter stick surface?

b. Press harder in the center of the meter stick surface. Now what does your finger feel? What happens to the meter stick surface?

c. Repeat parts a. and b. using only one meter stick as the surface. Does the surface bend more or less than it did when two meter sticks were used?

d. Do a similar set of investigations with 50-g, then 100-g, and finally 150-g masses placed in the center of the meter stick surface. Describe what happens.

e. What mechanism do you think might explain the ability of the meter sticks to react to an active force by applying a normal force? Is there any relationship to the mechanism you used to explain the phenomenon of tension?

f. Based on your observations with two meter sticks, what would happen to the bending of the surface if we used hundreds of meter sticks in a stack to resist a person's push or to hold up a mass?

g. Does a table or a wall bend noticeably if an active force is applied to it? What mechanism do you propose to explain how walls and tables can exert normal forces without bending noticeably? Do you think that the surface moves at all?

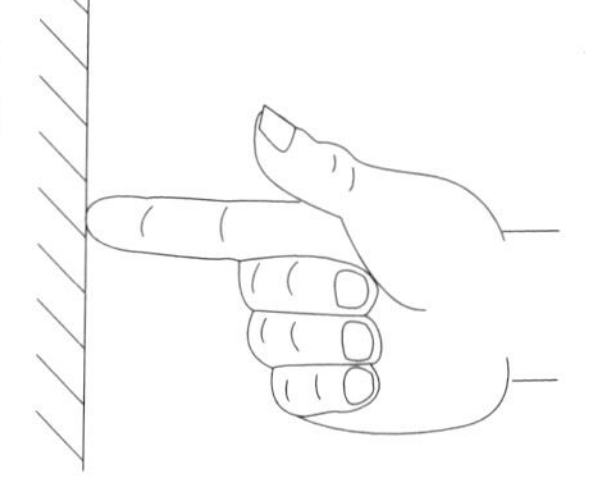

Fig. 7.18.

h. The diagram below shows a block sliding along a table near the surface of the earth at a constant velocity. According to Newton's first law, what is the net force on the block? In other words, what is the vector sum of all the forces on the block?

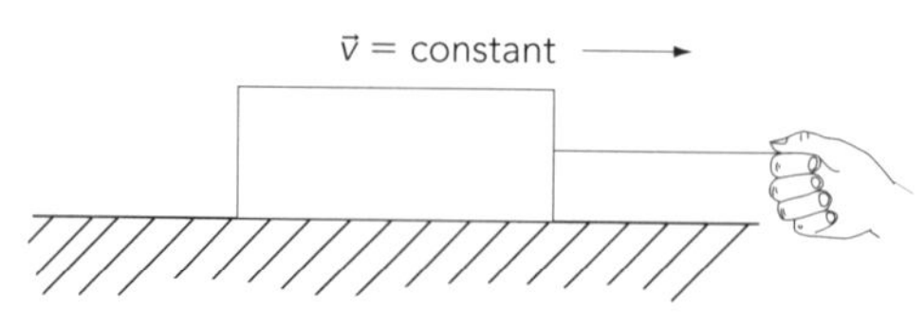

Fig. 7.19.

$$\Sigma\vec{F}=$$

i. The net force is made up of four mutually perpendicular forces. What are the two active forces? What are the two passive forces? In what direction does each one act? Draw a diagram indicating the direction of each of the forces.

7.11. GRAVITATIONAL FORCE ON A MASS ON AN INCLINE

Suppose that a cart of mass m is on an incline of angle θ and is held in place by an applied force as shown in the following diagram. If you know the angle of the incline and the magnitude and direction of the gravitational force vector, what do you predict the equation will be that describes the force component parallel to the plane? Perpendicular to the plane?

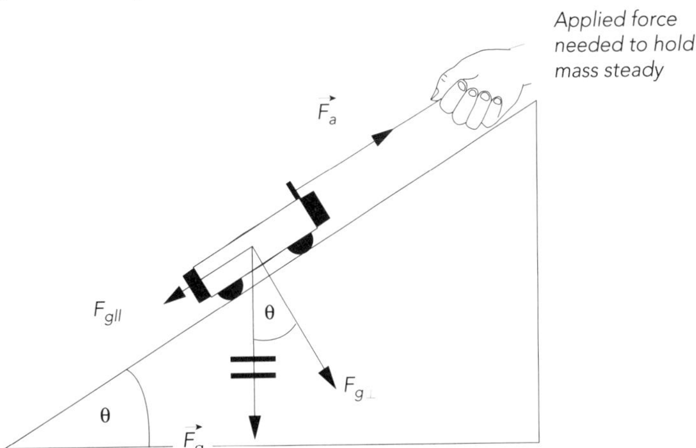

Fig. 7.20 A cart on an inclined plane. The vertical gravitational force is broken into two components—one perpendicular to the plane and one parallel to it.

In order to complete the following activity you will need:

- 1 ramp
- 1 low-friction cart
- 1 spring scale, 5 N (or more for a massive cart)
- 1 protractor

Recommended group size:	2	Interactive demo OK?:	N

7.11.1. Activity: Components of F_g on an Incline

a. The angle that the incline makes with the horizontal is the same as the angle between the perpendicular component of force and the F_g direction. Explain why.

b. Choose a coordinate system with the x-axis parallel to the plane with the positive direction up the plane. Using normal mathematical techniques for finding the components of a vector, find the values of $F_{g||}$ and $F_{g\perp}$ and as a function of the angle of the incline θ and of the magnitude of the gravitational force F_g.

c. What is the equation for the magnitude of the net normal force exerted on the cart by the surface of the incline? **Hint:** Use Newton's first law and the knowledge that the block is not moving in a direction perpendicular to the plane.

FRICTION AND APPLYING THE LAWS OF MOTION

7.12. PREDICTING AND MEASURING FRICTION FACTORS

If Newton's laws are to be used to describe the sliding of a block in *contact* with a flat surface, we must postulate the existence of a passive friction force that crops up to oppose the applied force. There are two kinds of friction forces: *static friction* and *kinetic* or *sliding friction,* which is the friction between surfaces in relative motion. We will concentrate on the study of kinetic friction for a sliding block. For this project you will have the following equipment available:

- 1 block w/ hook
- 1 metric spring scale (0–5 N)
- 4 masses, 200 g
- 1 electronic scale, 1 kg
- 4 flat surfaces (or more)

Optional: a computer-based laboratory system

- 1 force sensor
- 1 motion software

Recommended group size:	2	Interactive demo OK?:	N

7.12.1. Activity: Prediction of Friction Factors

a. List several parameters that might influence the magnitude of the kinetic friction force.

1. ______________________
2. ______________________
3. ______________________
4. ______________________
5. ______________________
6. ______________________

b. Pick one of the factors that interests you and describe how you might do an experiment to determine the effect of that factor on the magnitude of the friction force.

Measuring Kinetic Friction Factors

Let's determine what parameters actually influence the friction force. Students at each table can vary one parameter of interest to the class and determine the friction forces associated with it. Each group should analyze its results *using a computer graphing routine.* Enter the data points directly onto a computer, label each column, and produce an appropriate scatter plot.

7.12.2. Activity: Friction Data and Analysis

a. Describe the factor your group studied, create a data table for the friction force as a function of that factor, and then summarize your data and do a sketch of the scatter graph of your sliding friction force as a function of your factor (or place a computer printout of your graph in the space below).

b. If you didn't study the friction force as a function of mass, summarize the data taken by some of your classmates and include a graph of F_f vs. m below. What is the meaning of the slope of the graph?

c. Look up kinetic friction in the index of a textbook. Read about the coefficient of sliding friction, μ_k, and figure out how to determine μ_k from the data you have taken or might take. Using data provided by those groups that studied F_f as a function of m, calculate μ_k for the block sliding on one of the surfaces in the lab. Be sure to specify the two surfaces that were in contact.

7.13. THEORIES OF FRICTION

No material is perfectly "smooth and flat." Any surface when examined under a microscope is full of irregularities. It is usually assumed that sliding friction forces result from the rubbing of rough surfaces—that is, from the interlocking of surface bumps during the sliding process. How reasonable is this explanation for sliding friction? To make the observations described below you should have following equipment:

- 1 block with hook
- 1 metric spring scale (0–5 N)
- 4 or more clean surfaces, including at least one of acetate used for making overhead transparencies

Optional: Computer-based laboratory system with
- 1 force sensor
- 1 motion sensor
- 1 motion sofware

Recommended group size:	2	Interactive demo OK?:	N

7.13.1. Activity: What Surfaces Have High Friction?

a. Which kinds of surfaces do you think will have the most friction—rough ones or smooth ones? Why?

b. Make some qualitative observations to test your hypothesis by sliding the smooth wooden block on some smooth and rough surfaces. Make sure that at least one of these is a smooth sheet of glass or Plexiglas. Which situation has the most sliding friction with it? How does your observation compare with your prediction? Are you surprised?

Adhesion

The fact that smooth surfaces sometimes have more sliding friction associated with them than rough surfaces has led to the modern view that other factors such as *adhesion* (i.e., the attraction between molecules on sliding surfaces) also play a major role in friction. Predicting the coefficient of sliding friction for different types of surfaces is not always possible and there is much yet to be learned about the nature of the forces that govern sliding friction. Some authors of physics texts tend, incorrectly, to equate smooth surfaces with "frictionless ones" and to claim that the rubbing of rough surfaces is the primary cause of friction.*

* Some good references on friction are: W. H. Sherwood, "Work and Heat Transfer in the Presence of Sliding Friction," *Am. J. Phys.*, Vol. 52 (1984), pp. 1001–7, and D. Tabor , "Friction—The Present State of Our Understanding," *J. of Lubrication Technology*, V. 103 (1981), pp. 169–79.

7.14. FREE-BODY DIAGRAMS–PUTTING IT ALL TOGETHER

You have made observations that we hope led you to reconstruct Newton's three laws of motion for yourself. Here's a summary of these laws.

Summary of Newton's Laws

First: If the net force acting on an object is zero, its acceleration is zero.

$$\text{If } \Sigma\vec{F} = 0, \text{ then } \vec{a} = 0 \text{ so that } \vec{v} = \text{constant or } 0$$

Second: The net force on an object can be calculated by multiplying its mass times its acceleration.

$$\Sigma\vec{F} = m\vec{a}$$

Third: Any two objects that interact will exert forces on each other that are equal in magnitude and opposite in direction.

$$\vec{F}_{12} = -\vec{F}_{21}$$

These three laws are incredibly powerful. An understanding of them allows you to either: (1) use a complete knowledge of forces on a system of objects to predict motions in the system, or (2) identify the forces on a system of objects based on observations of its motions. In fact, you have already used a belief in Newton's laws to identify several active and passive "invisible" forces." The forces identified so far are shown below.

Force Notation

Active forces:

External forces (pushes or pulls)	$\vec{F}_{ext}$
Gravitational force	$\vec{F}_g = m\vec{a}_g$
Centripetal force	$\vec{F}_c = m\vec{a}_c = \frac{mv^2}{r}\,\hat{r}$

Passive forces:

Tension force	$\vec{F}_T$
Normal force	$\vec{F}_{norm}$
Static friction force	$\vec{F}_{f,\,stat} = \mu_{f,\,stat}\,\vec{F}_{norm}$
Kinetic friction force	$\vec{F}_{f,\,kin} = \mu_{f,\,kin}\,\vec{F}_{norm}$

We will eventually study other forces such as spring forces, air friction, electrical forces, and magnetic forces, but for now we will only deal with the forces summarized above.

Using Free-Body Diagrams to Calculate Forces and Predict Motions

The key to the effective application of Newton's laws is to identify and diagram all the forces acting on each object in a system of interest. The next step is to define a coordinate system and break the forces down into components to take advantage of the fact that if

$$\Sigma\vec{F} = m\vec{a} \quad \text{then} \quad \Sigma F_x = ma_x \quad \text{and} \quad \Sigma F_y = ma_y$$

A *free-body diagram* consists of a set of arrows representing all the forces on an object, but *not* the forces that the object exerts on other objects. To create a free-body diagram you should do as follows:

1. Draw arrows to represent all force acting on the object or objects in the system of interest.
2. Place the tail of each arrow at the point where the force acts on the object.
3. Point each arrow in the direction of the force it represents.
4. If possible, make the relative lengths of the arrows correspond to the magnitudes of the forces and label all the arrows using standard notation to show the type of force involved.
5. Choose and indicate a set of coordinate axes.

Important Note: The idea of using a single force vector to summarize external forces that act in the same direction is a useful simplification that is not realistic. For example, when a block rests on a table, we will say that the table exerts a normal force on the block. It is conventional to draw a single upward arrow at the point where the middle of the bottom surface of the block touches the table. This arrow actually represents the sum of all the smaller forces at each point where the block touches the table. This is shown in the following diagram.

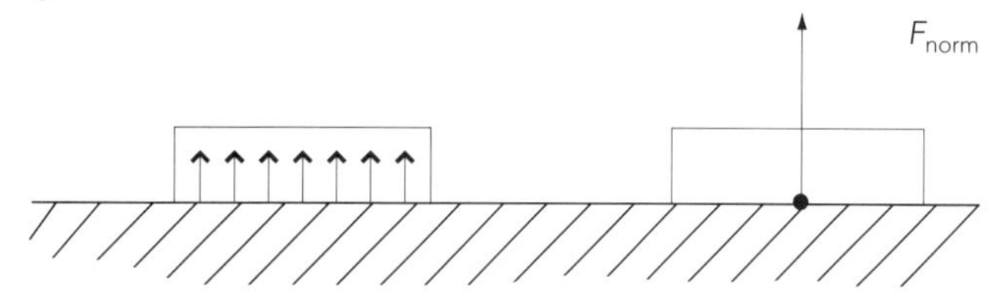

Fig. 7.21. For simplification, many small force vectors supporting the bottom of the block are replaced by a single large force vector acting through the center of the block.

A four wheeled cart will actually be supported in each of the four places where the wheels touch. The magnitude of the net normal force (F_N) is the sum of the individual normal forces ($F_N/4$).

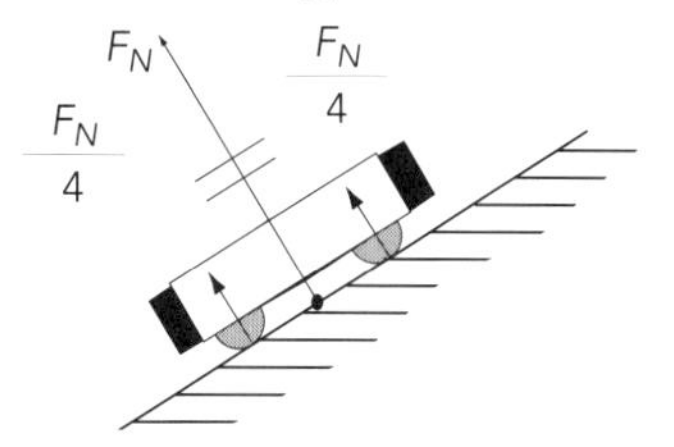

Fig. 7.22. Examples of how the net normal force can represent the sum of smaller normal forces.

An Example of a Free-Body Diagram

Consider a block of mass m sliding down a rough inclined plane as shown in the diagram below. It has three forces on it: (1) a gravitational force, (2) a normal force perpendicular to the surface of the plane, and (3) the friction force opposing its motion down the plane.

Since the block is sliding, it could either be moving at a constant velocity or with a constant acceleration. Thus, it is possible that the vector sum of forces on it is not equal to zero in some cases. For example, if there is accelerated motion, then:

$$\sum \vec{F} = \vec{F}_{\text{norm}} + \vec{F}_{f,\,kin} + \vec{F}_g = m\vec{a}$$

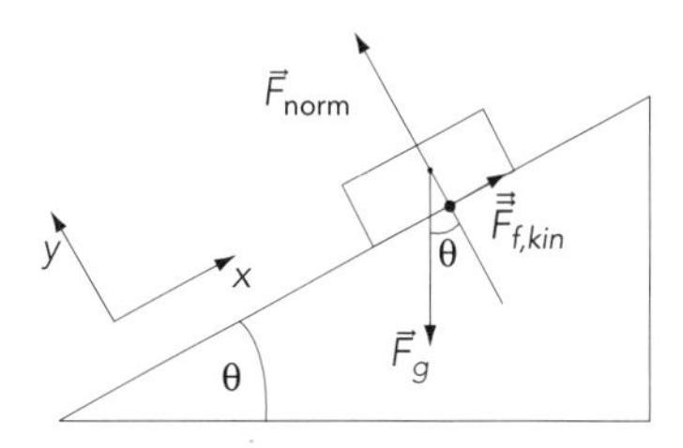

Fig. 7.23. An example of a free-body diagram for a block sliding down a ramp in the presence of sliding friction.

Based on the example in Figure 7.23, try your hand at drawing the free-body diagrams for the situations described below. In each case, write the equation for the vector sum of forces.

7.14.1. Activity: Some Free-Body Diagrams

a. A block slides freely down a frictionless inclined plane.

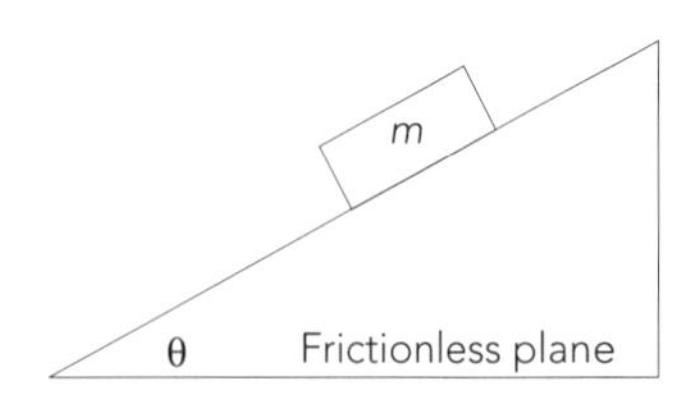

b. A block is on a plane with friction but is not moving due to static friction.

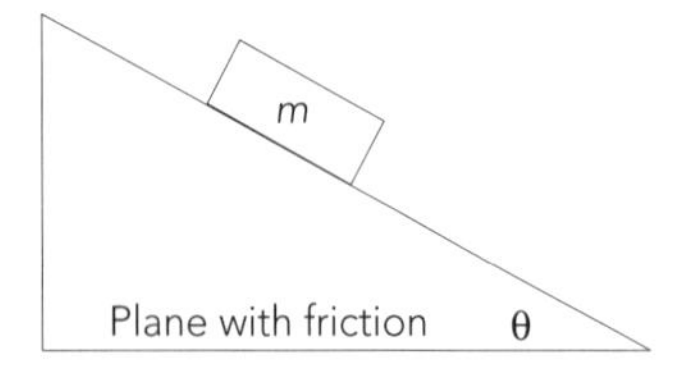

c. A block is on a plane with friction but the coefficient of friction between the block and the plane is small. The block is sliding down the plane at a constant velocity so that kinetic friction is acting.

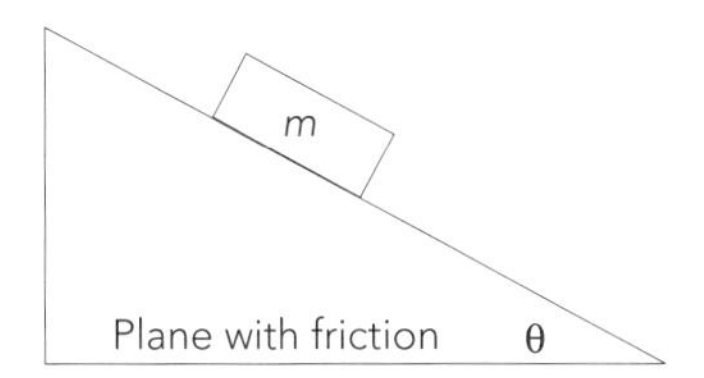

d. The block is on a frictionless plane but is attached to a hanging mass of $2m$ by one of our famous "massless" strings over a pulley. Construct free-body diagrams for the forces on the mass m and for the forces on the mass $2m$.

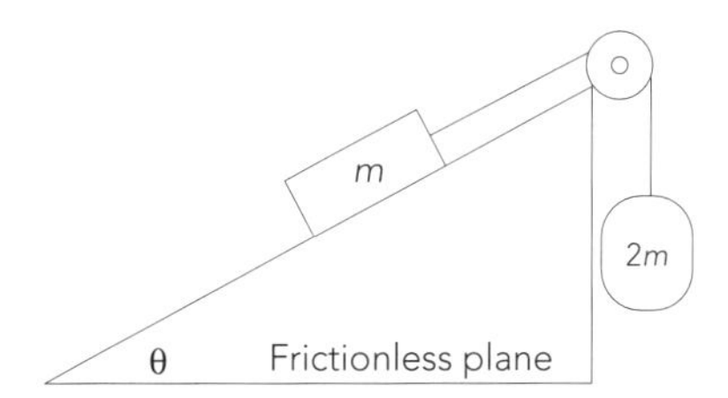

e. Describe what motions, if any, could result from the configuration in d?

7.15. BREAKING THE FORCES INTO COMPONENTS–AN EXAMPLE

Let's consider the case of the cart rolling down a ramp with a negligible amount of friction. Suppose the ramp makes a known angle with the horizontal. What will the acceleration of the cart be according to theory—that is, using Newton's laws? How will it compare to the magnitude of the gravitational acceleration a_g? In the activities in this section you will predict the acceleration, use theoretical equations to calculate the expected acceleration, and then measure the acceleration.

The free-body diagram and coordinate system chosen for analysis are shown in Fig. 7.24. By convention, if we use a symbol with a little arrow over it as we have for the gravitational force

$$\vec{F}_g$$

it represents a vector with both magnitude and direction. The absence of a little arrow over a vector quantity indicates that we are representing only the magnitude of the vector, which is always assumed to be positive. In the diagram below, F_g is a positive quantity; When taking components of the magnitude of the gravitational force, F_g, it may be necessary to introduce a negative sign to indicate whether the component points in a positive or negative direction.

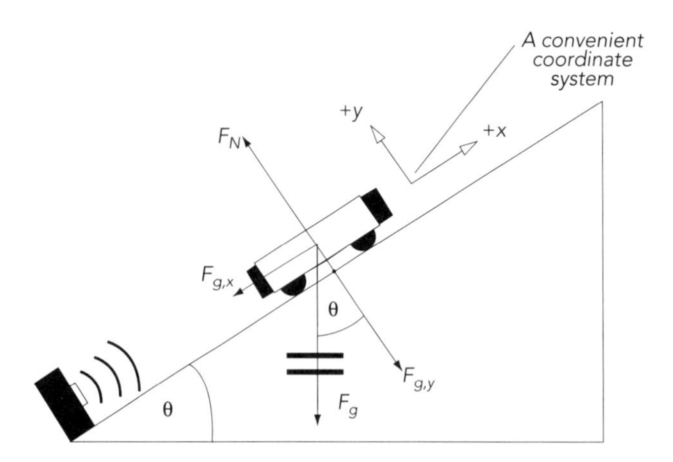

Fig. 7.24. A cart with no external applied forces on it such as a human hand.

For the activities that follow you will need:

- 1 cart
- 1 ramp
- 2 textbooks (to prop up the ramp)
- 1 meter stick
- 1 protractor
- 1 computer-based laboratory system
- 1 ultrasonic motion sensor
- 1 motion software

Recommended group size:	All	Interactive demo OK?:	Y

7.15.1. Activity: Predicting Cart Motion on a Ramp

a. If the ramp is tilted at some angle with respect to the horizontal that is greater than zero and less than 90 degrees, what kind of motion do you expect the cart to have? Constant acceleration? Constant velocity? Something else?

b. Consider the x-components of acceleration and velocity as shown. Using the axes below, sketch the shapes of the velocity vs. time and of the acceleration vs. time graphs along the x-direction. Assume that someone has given the cart an initial shove up the ramp just before the clock is started and that a motion sensor is at the bottom of the ramp.

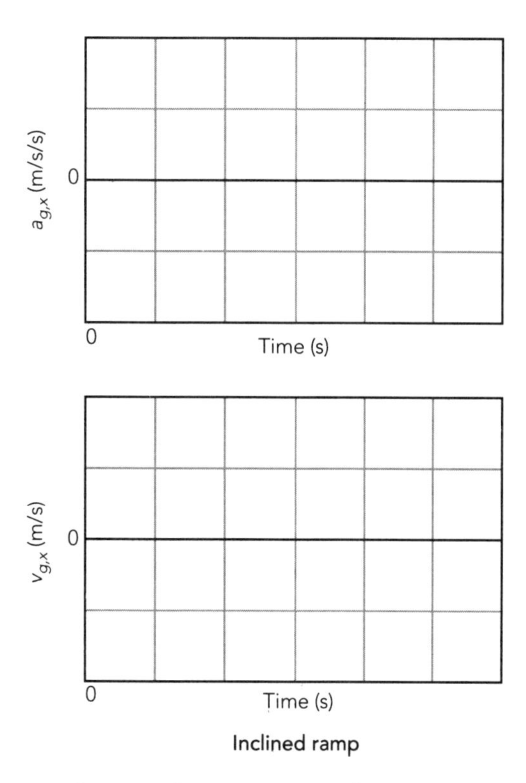

Inclined ramp

c. With respect to the coordinate system shown in Figure 7.24, will the x-component of cart acceleration be zero all of the time, only when the cart is turning around, or none of the time? Explain.

d. With respect to the coordinate system shown in Figure 7.24, will the y-component of cart acceleration be zero all of the time, only when the cart is turning around, or none of the time? Explain.

Deriving Equations for the Acceleration Components

Let's use trigonometry to find the components of the forces shown in Figure 7.24. If the gravitational force has a magnitude given by F_g, then we can break it into components as follows.

$$F_{g,x} = -F_g \sin(\theta)$$
$$F_{g,y} = -F_g \cos(\theta)$$

There is no motion perpendicular to the surface, so the net normal force must be equal in magnitude and opposite in direction to the y-component of the gravitational force. Thus,

$$\Sigma F_y = 0 = F_{norm} + F_{g,y}$$
$$F_{norm} = -F_{g,y} = F_g \cos(\theta)$$

There is no balancing force for the x-component of F_g, so, according to Newton's second law,

$$\Sigma F_x = ma_x = -F_g \sin(\theta) = -ma_g \sin(\theta)$$
$$a_x = -a_g \sin(\theta)$$

Calculating and Observing Acceleration on an Incline

The next step is to set up the ramp an an angle, calculate the theoretical acceleration for that angle, and check the validity of the calculation experimentally.

7.15.2. Activity: Calculating and Measuring Acceleration

a. Prop the ramp up and figure out a way to determine the angle it makes with the horizontal without using a protractor. Explain any measurements and calculations you used to determine the angle.

b. Calculate the theoretical value to the cart acceleration for an incline having the angle you determined.

c. Practice giving the cart an initial velocity up the ramp and then watch its motion. Describe the motion you observe using words.

d. Use the computer-based laboratory system and motion sensor to measure the x-component of the acceleration of the cart as it travels up and down the ramp. Sketch graphs of the x-components of velocity vs. time and acceleration vs. time below (or affix a computer printout of your results). Please include the scale for time, velocity, and acceleration on each graph.

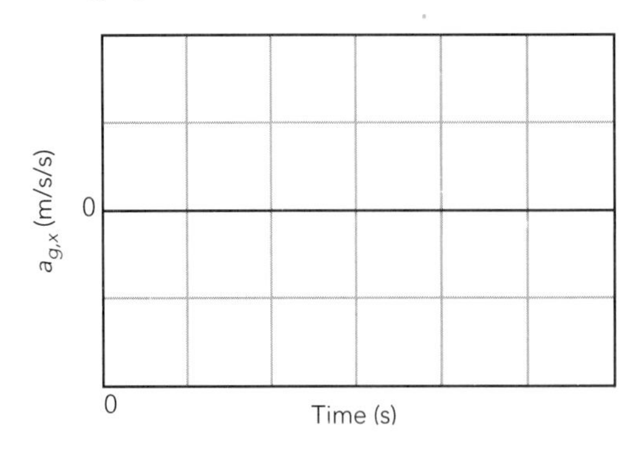

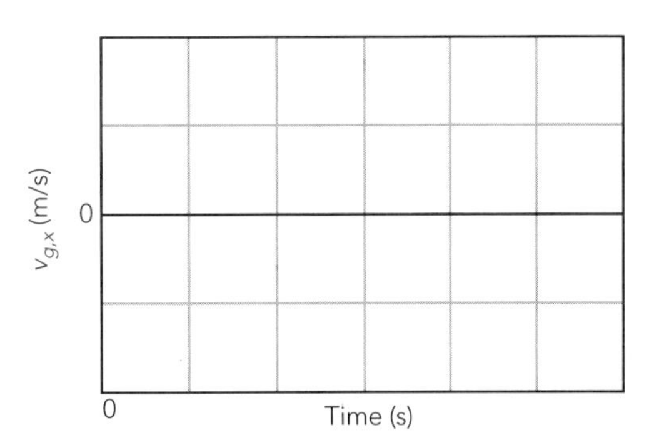

Cart traveling on an inclined ramp

e. How do the shapes of the graphs compare to those you predicted?

f. Is the acceleration constant within the limits of experimental uncertainty? What is its average value as the cart goes up and down the ramp?

g. How does the measured acceleration compare with the value you calculated theoretically?

APPENDIX A
COMPUTER SPREADSHEETS AND GRAPHS

USING MICROSOFT® EXCEL 5.0 WITH THE CUSTOM WORKSHOP PHYSICS TOOLS (WP TOOLS)

OVERVIEW

This Workshop Physics *Activity Guide* includes many spreadsheet and graphing exercises for data analysis and numerical problem solving, as well as analytic and numerical modeling of data. Several spreadsheet software packages include graphing routines that can be used for these exercises. Some, such as Claris Works, are relatively inexpensive and have modest hardware requirements. Others, like Microsoft Excel 4.0 and 5.0, have a full array of features, however, require more powerful hardware. The spreadsheet and graphing package you choose will depend upon the hardware you have available and the capabilities required for the Workshop Physics activities you intend to perform.

In these appendices we have chosen to describe the use of Excel 5.0 complemented by an Excel add-in called WPtools that has been developed for the Workshop Physics program. Excel 5.0 has a wide range of features and is sufficient for use with all of the Workshop Physics spreadsheet activities. However, other software packages that also can be used include Claris Works, Delta Graph, Lotus 1-2-3, and KaleidaGraph.

Microsoft Excel® 5.0 is a software package that combines spreadsheet accounting, graphing, and database functions into one program. Starting with a spreadsheet, you can enter, modify, and analyze data. You can add text to the spreadsheet to write notes about the data that was collected. Finally, the data can be graphed. The WPtools Excel add-in adds some new features to Excel 5.0 and extends others to make the Workshop Physics activities easier. At the same time, the WPtools remain general purpose and useful in a wide range of other applications.

Included in this appendix are detailed steps for using some Microsoft® Excel features on Macintosh or PC-compatible computers with Microsoft® Windows. You will learn many of these steps readily; others will take more time. Before doing much work with the spreadsheet and graphing software, familiarize yourself with the features of your computer (the Macintosh or PC compatible/Windows) by running the tutorial and reference programs provided with your computer. In this appendix we discuss: (1) basic Macintosh/Windows operations; and (2) how to use Excel for creating spreadsheets and graphing data, including the use of the custom WPtools.

GLOSSARY OF TERMS

The following are common terms that will be used throughout this and other appendices. These terms will make more sense as you run the computer-based tutorial programs for your computer.

Click: To click on an object, move the mouse until the arrow (or another form of the cursor) rests at the point you want. Then press and release the button on the mouse.

Command key (Macintosh): The key to the left of the space bar — with the apple and flower-like symbol(⌘)— is called the command key. It is often used in conjunction with other keys to choose menu items.

Control Key (Windows): This is the key below the shift key on each side of the keyboard. It is often used in conjunction with other keys to choose menu items.

Desktop: The desktop is the screen that displays objects such as icons and windows that can be opened for viewing.

Double click: A double-click is often used for opening an application or file. Rest the cursor at the point of interest, then rapidly click the mouse button twice in succession.

Drag: To drag, center the cursor on the object, push and hold the mouse button down, then slide the mouse to move the cursor. When the desired end location is reached, release the mouse button. If you drag past the desired end location, do not release the button; just move it back to the desired location, then release it.

Icon: An icon is a small picture that represents programs, sets of data, texts, graphs, and other objects.

Menu: The menu bar holds a list of commands and appears across the top of the screen on the Macintosh and across the top of each window in Windows. To select a command, click on a specific menu title, drag down the menu until the command you want is highlighted, then release the mouse button. Many of the commands listed in a menu can also be issued from the keyboard by holding down the command key for Macintosh or the control key for Windows and pressing the letter which corresponds to the command that you want to access. Letters are listed next to the commands in the menu.

Select: To select something, such as an icon or a spreadsheet cell, move the cursor until it rests at the point of interest and click on the mouse button.

Window: Windows are rectangles on the desktop containing icons, application programs, the contents of a file or even other windows. Windows can be resized, moved around, and overlapped with other windows.

ABOUT EXCEL 5.0 AND THE WPtools

Excel is a popular spreadsheet and graphing program. A spreadsheet is made up of numbered rows and lettered columns that form cells. A cell is the intersection of a row and a column and has a letter and a number associated with it. Several types of information can be entered into a cell, such as words, numbers, or even mathematical formulas. You can use a spreadsheet to do many different things. Spreadsheets can be used to balance your checkbook, do your taxes, or analyze physics data.

A custom set of tools for use with Excel 5.0 has been created for use with the Workshop Physics and Workshop Mathematics courses at Dickinson College. Some of the instructions included in this appendix will help you use any installation of Excel 5.0. Other instructions explain how to use the custom tools that are available by installing the custom WPtools add-in provided with your Activity Guide. A similar set of custom tools for use with Excel 4.0 is also provided with the Activity Guide.

OPENING AN EXCEL SPREADSHEET

To start working with a blank Excel spreadsheet, you need to:

a. Turn on the computer.

b. Launch the Excel 5.0 program by double-clicking on its icon.

c. After a brief time the Excel program will be loaded to the active memory of your computer and an empty workbook like that shown in figure A.1 appears.

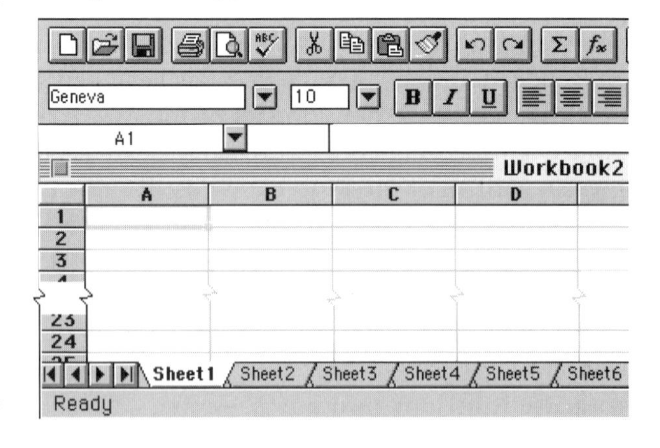

Fig. A.1.

HOW TO USE THE SPREADSHEET– AN EXAMPLE

Spreadsheets are very useful for analyzing and graphing scientific data. For example, if we want to determine someone's average pitching speed, we could use a spreadsheet to analyze data on the distance a baseball travels in a certain amount of time. In such an analysis we want to calculate the speed of each pitch and the average speed of all the pitches. We will assume that you have collected data on distances and times of travel for a pitched baseball.

Entering Information and Data into Cells

Now that you have your data, you need to enter them into the spreadsheet. Titles for each column can be entered in the top cells to help in your analysis.

1. Alphabetic Titles (and text in general)
Use the mouse to select the cell in which you want to put the first title (i.e., cell A1); a selected cell is indicated by a bold border. Type in the title, Distance (m), as shown in Figure A.2. After you type the title, select the next cell into which you want to enter information (i.e., cell B2). By selecting a new cell the information you just typed is entered into its cell. Pressing the Return key (Macintosh) / Enter key (Windows), the Tab key, or any Arrow key also enters the information. Add the remaining titles to the spreadsheet in the cells as indicated below:

Distance (m)	in cell A1
Time (s)	in cells B1,C1,and D1
Average time (s)	in cell E1

Your spreadsheet should now look like the one shown in Figure A.2.

	A	B	C	D	E
1	Distance (m)	Time (s)	Time (s)	Time (s)	Average Time (s)
2					
3					

Fig. A.2.

2. Entering Data

Suppose the times you recorded in seconds for the three throws are:

for 10m, t = 0.45s; 0.50s; 0.47s
for 15m, t = 0.76s; 0.92s; 0.85s
for 20m, t = 1.02s; 0.97s; 0.95s
for 30m, t = 1.46s; 1.44s; 1.34s

To enter the data, select the cell you want to put the data in and type the number. You can change which cell is selected by using the mouse or the arrow keys on the keyboard. Your spreadsheet should now look like that shown in Figure A.3.

	A	B	C	D	E
1	Distance (m)	Time (s)	Time (s)	Time (s)	Average Time (s)
2	10	0.45	0.5	0.47	
3	15	0.76	0.92	0.48	
4	20	1.02	0.97	0.95	
5	25	1.46	1.44	1.34	
6					

Fig. A.3.

3. Entering Formulas
To enter the formula that will calculate the average value of your times for each distance, follow these steps:

a. Select the appropriate cell for the average time (for the first set corresponding to a distance of 10 meters, the location is cell E2).

b. Type: $=(B2+C2+D2)/3$. The screen should look approximately like that shown in Figure A.4.

c. Hit the return key or select the checkmark next to the formula bar. The value for the average of the three points should appear in the cell you selected.

E2 =(B2+C2+D2)/3

Appendix A.xlw

	A	B	C	D	E
1	Distance (m)	Time (s)	Time (s)	Time (s)	Average Time (s)
2	10	0.45	0.5	0.47	=(B2+C2+D2)/3
3	15	0.76	0.92	0.48	
4	20	1.02	0.97	0.95	
5	25	1.46	1.44	1.34	
6					

Fig. A.4.

To duplicate the formula into the three cells below it, drag the cursor from E2 through cells E3, E4 and E5 so that they are all highlighted. Next move the cursor to the *Edit* menu and drag the selector down to *Fill*; when *Fill* is selected, drag the cursor right to get a pop-up menu. When the pop-up menu appears, drag the cursor to the *Down* option and release the button. The formula will automatically be copied into the cells. Or, if you wish, you can use the Command key (Macintosh) / Control key (Windows) and D key instead of using the *Edit* menu. Any time you see a symbol next to a command in a menu, like the D (Macintosh) or CTRL+D (Windows) that appears next to the *Down* command, it means that the same function will be performed if you press ⌘ and the indicated key simultaneously.

If you know how many significant figures to use, you should format your data cells to display the number of significant figures you want. This is done by first selecting the cells you want to format. Then open the Format menu and drag the cursor to the *Cells . . .* option and release the button (or use ⌘1/ CTRL+1). A dialog box titled Format Cells will appear. Click on the tab labeled "Number" at the top of the dialog box. Select the form in which you want your numbers to appear and select OK. If the form you want isn't listed (five decimal places, for example) then type it into the box labeled "Code:". If needed, you can format the cells for scientific notation. For example, to use scientific notation with three significant figures, you should type the symbol "0.00E + 00 into the "Code:" box.

Your spreadsheet should now look like the one in Figure A.5.

E2 =(B2+C2+D2)/3

Appendix A.xlw

	A	B	C	D	E
1	Distance (m)	Time (s)	Time (s)	Time (s)	Average Time (s)
2	10.0	0.45	0.50	0.47	0.47
3	15.0	0.76	0.92	0.48	0.72
4	20.0	1.02	0.97	0.95	0.98
5	25.0	1.46	1.44	1.34	1.41
6					

Fig. A.5.

4. *Playing Around*

Try exploring different menus to find out how some of the commands work. For a complete description, ask the instructor for a Microsoft Excel manual. You might try deleting some data; to do this, select a cell holding data, and then press "delete" and then "return" for Macintosh, or press "Back Space" and then "Enter" for Windows. Watch what happens to the average when one of the data points is erased. Be sure to re-insert any data you deleted before saving the file! On the Macintosh, you can get more explanations on some of the buttons and commands in Excel by clicking on the help menu ⓘ on the right side of the menu bar. From the menu that appears, select *Show Balloons.* Now, balloons with brief explanations will show up for things at which you point the cursor. To turn off the balloons, go back to the help menu and select *Hide Balloons.* For both Windows and the Macintosh, a small yellow box with the name of a tool will appear when the cursor is pointed at a tool.

5. *Using Functions Instead of Formulas*

Many of the common mathematical functions and formulas are available for use with the spreadsheet. For example, instead of entering the formula =(B2+C2+D2)/3 in cell E2, you could have used the "Average" function. To do this:

a. Select cell E2

b. Type: =*average* (

c. Next select all the data points to be included in the average. (In this case, B2 through D2). Move the arrow to cell B2 and drag it across the row to D2.

d. Type: **)** so that the formula in the formula bar of the screen is shown as "= AVERAGE(B2:D2)"

e. Hit the return key. The value for the average of the three times should again appear in cell E2.

There are many other functions available, including sin, cos, and StDev (standard deviation). A list of functions can be obtained by opening the *Insert* menu and selecting *Function . . . .* A dialog box titled "Function Wizard" will appear. Select the function you want to use and click the next button. A new dialog box will appear that allows you to specify the arguments to the function. The Function Wizard is a fancy feature that will save you time in remembering the format of each function. More details about all of the functions are contained in the Microsoft®Excel Manual and in the electronic help file. On the Macintosh, the help file is opened by choosing *Microsoft Excel Help . . .* from the help ⓘ menu. In Windows, the help file is accessed by choosing *Contents . . .* from the *Help* menu.

6. *Using Other Spreadsheet Features*

Two other features that are very useful in physics computations include formatting data and changing the width of columns.

a. *Formatting:* The *Format* menu allows you to change the appearance of your spreadsheet. To format cells, select the cells to be formatted and choose *Cells . . .* from the *Format* menu. Each of the tabs at the top of the dialog box allows you to alter different properties of the selected cells. You should experiment with the different properties and watch what happens in each case.

b. *Column Width:* If a column is too narrow to hold your values or wider than needed, you can change its width. Move the arrow to the very top of a column, point to the dividing line between the columns, click the mouse, and drag the divider line to the left or right. If you double-click on the divider line, the column will automatically change width to fit the widest cell in the column. There is also a *Width . . .* option on the *Column* pop-up menu on the *Format* menu.

7. *Saving a Spreadsheet*
In order to SAVE a spreadsheet for future use, you should:

a. Insert a preformatted 3.5″ data disk into the floppy disk drive in your computer.

b. Choose Save from the File menu.

c. A dialog box will appear. First observe the titles of the files, programs, or folders (Macintosh) / directories (Windows) displayed in the scrolling windows on the left side of the dialog box. To save your spreadsheet on your own floppy disk: For the Macintosh, click on the "Desktop" button and select your disk from the scrolling box on the left side of the dialog box. For Windows, click on the arrow at the right of the box labeled Drives and select the floppy disk drive. Now your file can be saved on your floppy disk.

d. In the box labeled "File Name" type the name you wish to give your file (e.g., "Pitch"), and then click the "Save" button (Macintosh) / "OK" button (Windows).

Your spreadsheet is now saved on your disk. For the next part of the exercise, you need to have the class data in your spreadsheet. Close your file by selecting the *Close* command from the *File* menu. A file may also be closed on the Macintosh by clicking in the small rectangular box in the upper left-hand corner of the Excel window or in Windows by double-clicking on the small square box to the left of the *File* menu.

Open a new Excel workbook by selecting *New* from the *File* menu.

The easiest way to format the class data that you will be working with is to have the average times for each distance in their own column. The columns are then labeled with the distance as shown in Figure A.6. Putting the data in rows as you did for your own data makes it harder to drag over the cells to find the average. Title the columns and enter the class data in your spreadsheet.

Save the spreadsheet in a file on your floppy disk by selecting *Save As . . .* from the *File* menu and following the provided instructions for naming and saving the file. Try to think of short but descriptive names for your files such as "SampleSS."

Note: This is just another way of saving your spreadsheet.

HOW TO SORT DATA USING EXCEL

Now that you know some Excel spreadsheet basics, it is time to learn several more complicated features. One feature that will come in handy is sorting selected information by the contents of a row or column. Text information can be alphabetized and numbers can be put in order from lowest to highest or vice versa. For example, suppose you have analyzed pitching data for several individuals and you want to rank people according to how fast they pitch. To do this:

a. Select all the data that you want to sort.

Note: Any cells not selected will not be sorted. If you sort cells that contain equations referring to cells that are not sorted, your sheet will become mixed up.

Figure A.6 shows how the data should be selected. Notice that the headings for each column are not selected. Since you are sorting by rows, the columns will not change, and the titles do not need to be selected.

	A	B	C	D
1	Name	Distance (m)	Average t (s)	Average speed (m/s)
2	Nicoli Tesla	15	0.72	20.8
3	Marie Curie	15	0.64	23.4
4	Albert Einstein	15	0.67	22.4
5	Issac Newton	15	0.56	26.8
6				

Fig. A.6.

b. Once the data is selected, go to the menu bar, choose the *Data* menu, and drag the cursor down to *Sort. . . .* A dialog box similar to the one shown in Figure A.7 will appear.

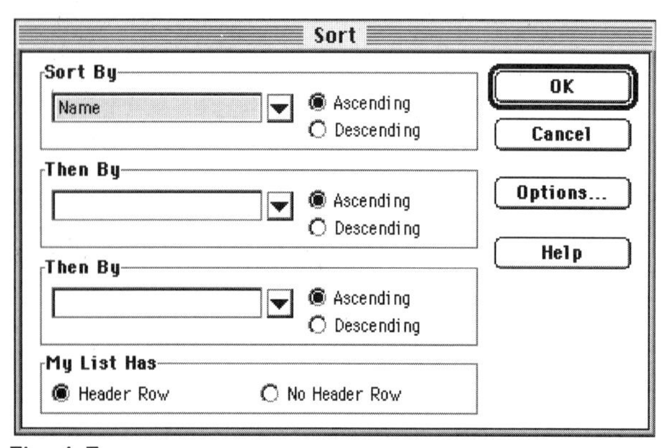

Fig. A.7.

The only part of the sort dialog that you will need to deal with is the *Sort By* part. Excel will sort the selected data according to the cell that is listed in the *Sort By* part of the "Sort dialog." It automatically lists the column label of the left-most column of the selection, which in this case is "Name."

c. Click the down arrow to the right of the text box in the Sort By area and select the column which contains the desired sort key. In this case it would be "Average speed (m/s)". Click OK to confirm your selection. The selected data will be sorted by the average speed values in column D.

The sorted data is shown in Figure A.8. Notice that Albert Einstein and Marie Curie have changed positions and that the average speeds are in ascending order.

Appendix A.xlw

	A	B	C	D
1	Name	Distance (m)	Average t (s)	Average speed (m/s)
2	Nicoli Tesla	15	0.72	20.8
3	Albert Einstein	15	0.67	22.4
4	Marie Curie	15	0.64	23.4
5	Issac Newton	15	0.56	26.8

Fig. A.8.

HOW TO USE THE EXCEL *CHART WIZARD* FOR GRAPHING

Excel allows you to graph data you have entered into a spreadsheet. Once you've gotten the hang of graphing, spend some time exploring Excel's many features.

To Graph Data

We will use the sample baseball pitching data to demonstrate how to graph using Excel. Suppose you want to plot the distance a baseball travels as a function of its average time-of-flight.

a. To use the Excel *Chart Wizard* to make a graph, you must organize your data properly in columns. The x-data (i.e., the horizontal axis data) should be placed in a column to the left of the column (or columns) of y-data (i.e., the vertical axis data). Column headers with variable names and units may be placed at the top of each column to denote what variable is being plotted. In this case, the x-data represents average time and the y-data the distance the ball has traveled in that time.

b. Once you have the data columns you want to graph next to each other, highlight them as shown in Figure A.9. (If you highlight the column headers, the dependent variable column labels will be used as series labels on your graph.) Since we want to know how far the baseball has traveled as a function of time, the time axis is horizontal and the distance axis is vertical.

Note: Excel will always take the leftmost highlighted column to be the x-data. If your x-data is to the right of your y-data, you will have to move the columns around before graphing.

	A	B	C	D	E
1	Time (s)	Time (s)	Time (s)	Average Time (s)	Distance (m)
2	0.45	0.50	0.47	0.47	10.0
3	0.76	0.92	0.48	0.72	15.0
4	1.02	0.97	0.95	0.98	20.0
5	1.46	1.44	1.34	1.41	25.0

Fig. A.9.

c. Click on the *Chart Wizard* icon located on the tool bar in the top right part of the screen. You will then be prompted to do the following:

1. Select an area in the spreadsheet window for your graph. Do this by positioning the cursor, which will be a + symbol, at the upper left-hand corner of your desired graph location (which must be somewhere on your worksheet). Press the mouse button and drag the cursor to the lower right-hand corner of your graph-to-be. (A rectangle will appear while you are dragging to indicate the region you have chosen.)

2. When the *Chart Wizard* dialog box appears,

the selected cells will be listed as the data range. The data range specifies the data from which the chart will be created. If you highlighted the correct data before choosing the *Chart Wizard*, you need only click on the *Next* button; otherwise use this opportunity to correct the data range.

3. Choose the type of graph you want and click on the *Next* button. In general, you will want to choose an *xy* scatter graph.

4. Choose the graph format you want and click on the *Next* button. For most physics applications, you will wish to choose scatter graph 1—unconnected dots on a non-logarithmic scale.

5. Indicate the method by which *Chart Wizard* should interpret your data table. If your data series are in columns and you included column headers as the top row of your highlighted data, you need only click on the *Next* button.

6. Enter any text you may wish to use as a graph title or axis labels for your graph and click on the *Finish* button.

At this point, Excel will create a graph using the data range specified above in step 2.

d. You can adjust the size of the graph by using the mouse. Place the cursor over one of the small dark squares, called handles, on the edge of the graph. Drag the handle to resize the graph frame. If the graph handles are not visible click once on the graph.

e. To activate a chart, double-click on it. The thin black border will be replaced by a thick border indicating that the chart is active. An active chart appears as shown in Figure A.10.

Using various menu options, you may now change the way your graph appears on the screen.

f. You can save your spreadsheet as explained before. If you just want to print a full-page copy of the graph, activate the chart and choose *Print . . .* from the *File* menu.

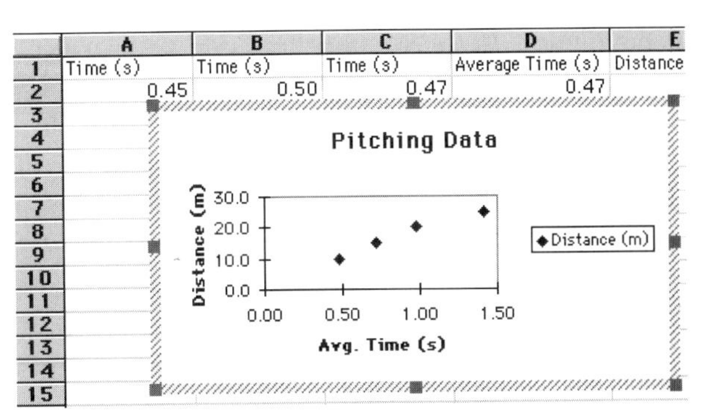

Fig. A.10.

Dynamic Linking — A Nice Feature!

If you think you might add more data to your spreadsheet at a later date, highlight extra cells when you first make the graph but leave them blank. Graph the data you do have plus the blank cells. When you later come back and place data in the blank cells you had included, the new data points will be graphed automatically. Also, if you change the number in a cell, the corresponding point on the graph will change automatically.

Play Around

This is just a brief introduction to the graphing capabilities of Excel. There are many interesting things you can do to the graph. You can change the font, the numbering of the axes, the types of graphs, the shape of the data points, and many other things, including adding text windows and pictures. Experiment with Excel a bit and get to know some of its unique capabilities.

USING THE WPTOOLS CUSTOM GRAPHING TOOL

These instructions explain how to graph using a special Scatter Plot tool which is part of the WPtools add-in package developed at Dickinson College. These instructions assume a default installation of WPtools. If the WPtools preferences have been modified, then the results of the operations described may be different from the figures shown. The Scatter Plot tool allows you to create a scatter plot including a title, axis labels, and legend in a single step. This tool is also more flexible than Excel's *Chart Wizard* for creating scatter plots. The columns of data do not have to be right next to each other, and the independent variable may be anywhere in relation to the dependent variable. However, there are some restrictions for formatting your data. A label should be entered above each column of data. The label above the independent variable column becomes the horizontal axis label. If there is a single dependent variable column, then its label becomes the vertical axis label. If there are multiple dependent variable columns, their labels appear in the plot legend, and the vertical axis is not labeled. A plot title which will appear at the top of the plot may be entered in the cell above the dependent variable column label. One possible format for a data table is shown in Figure A.11.

	A	B	C	D
1	Plot Title			
2	X-Axis Label	Dep. Var. 1	Dep. Var. 2	etc...
3	0.0	1.0	2.0	
4	1.0	2.0	4.0	
5	2.0	3.0	6.0	
6	3.0	4.0	8.0	
7	4.0	5.0	10.0	
8	5.0	6.0	12.0	

Fig. A.11.

After entering data, select the data you wish to plot as shown in Figure A.11. You may either select the graph title and the series labels, or not; they will nonetheless appear on your final graph.

After selecting the desired data, click on the Scatter Plot tool from the WP Standard toolbar or select *Scatter Plot* from the *WPtools* menu. The tool will create a graph without further prompting. (After it has been created, you can alter the size of the graph and make other changes to it in the normal fashion.) If you were to enter data as shown in Figure A.11, the custom Scatter Plot tool would create the graph shown in Figure A.12.

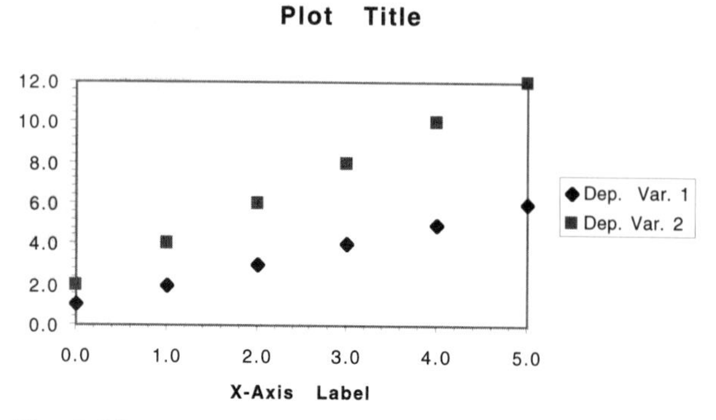

Fig. A.12.

USING THE WPTOOLS CUSTOM SCATTER PLOT TOOL IF THE COLUMNS ARE NOT NEXT TO EACH OTHER

a. To use the custom Scatter Plot tool, you must organize your data in columns. The x-data (i.e., the horizontal axis data) should be highlighted first (that is, the 1, 2, 3, 4, data). The first y-data column should be highlighted next by holding the command key (Macintosh) / Control key (Windows) while selecting the data. Additional y-data columns can now be selected using the same method (that is, the 10, 15, 23, 45 data). The y-data columns can be anywhere on the spreadsheet; they do not have to be to the right of the x-axis data.

b. If you want the axes to be labeled automatically, column headers with variable names and units must be placed at the top of each column to denote which variable is being plotted. The Plot Title must be placed above the x-data column label as shown in Figure A.13.

	A	B	C
1			Plot Title
2			X-Axis Label
3	Dep. Var. 1		1.0
4	-4.0		2.0
5	-1.0		3.0
6	4.0		4.0
7	11.0		5.0
8	20.0		

Fig. A.13.

c. Once the data are properly highlighted, click on the Scatter Plot tool from the WP Standard toolbar or select *Scatter Plot* from the *WPtools* menu.

If you made the selection properly using the data shown in Figure A.13, your graph will look like the one shown in Figure A.14.

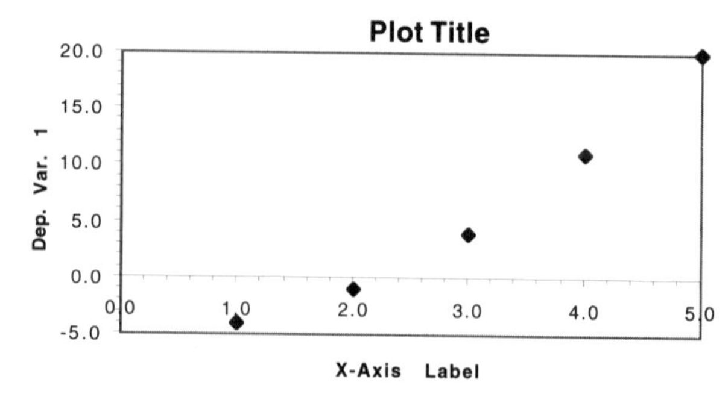

Fig. A.14.

PRINTING A SPREADSHEET OR GRAPH

In a classroom or laboratory setting there are often many people using the same printer. Thus you should always put your name and date at the top of each spreadsheet or graph (along with the graph title) before attempting to print. This will prevent your work from getting mixed up with that of other students.

To Print

a. Make sure that the printer is turned on and has had time to warm up.

b. When you have finished the spreadsheet or graph and have put your name and the date at the top, select *Print . . .* from the *File* menu.

c. When the box containing print options appears on the screen, select those you want and then click on OK. Your document will be printed in a minute or so.

d. If you wish to print another graph or spreadsheet, close the file you just printed. Then choose *Open . . .* from the *File* menu. When the Open window appears, open the file you want to print next by double-clicking on its title. Then proceed with the Print instructions above.

TO SHUT THINGS DOWN

Before quitting your work on the computer, you should do the following:

a. Save your file by selecting *Save.* If you want to give your file a new name, select *Save As . . .* from the *File* menu.

b. Quit the program you are using by selecting *Quit* (Macintosh) or *Exit* (Windows) from the *File* menu.

c. Shut the computer down using the appropriate procedure for your computer.

APPENDIX B
COMPUTER-BASED LABORATORY SOFTWARE AND HARDWARE*

WHAT IS A COMPUTER-BASED LABORATORY SYSTEM?

In most modern experimental laboratories, electronic sensors are used to collect data automatically. When a sensor is attached to a computer, a very powerful data collection, analysis, and display system is created. Computers coupled with appropriate software packages are capable of analyzing signals and displaying them on the screen in easily understood forms. Using these capabilities, a "real time" picture of the data in a symbolic representation, such as a graph, can be obtained.

The use of *real world* data and graphical representations provides an immediate picture of how a physical quantity such as an object's position or temperature changes over time. This picture leads to a better understanding of the data's significance. A system for the capture and display of scientific data, consisting of a microcomputer, an electronic interface, software, and sensors, is often called a *m̲icrocomputer-b̲ased l̲aboratory* system. MBL is a commonly used nickname for such a system. Another common name is simply *computer-based laboratory* system. (CBL is not used as a nickname because a recent new data recording system that works with a programmable calculator instead of a computer has been dubbed "CBL.") In this appendix all of these systems will be referred to collectively as Computer-Based Laboratory systems.

At present there are a number of Computer-Based Laboratory systems with electronic interface devices, associated sensors, and software that allow either a Macintosh or PC-compatible computer to be used as a laboratory instrument. The electronic interfaces transform signals from the sensors into forms which the computer accepts as input. In this appendix we will briefly describe three Computer-Based Laboratory systems, each of which uses a distinct electronic interface: The *Universal Laboratory Interface* (ULI) distributed by Vernier Software; the *Signal Interface II* (SI2) distributed by PASCO scientific; and the *Multipurpose Laboratory Interface* (MPLI) distributed by Vernier Software.

The ULI is designed for use with properly outfitted Macintosh or PC-compatible computers. The ULI information presented here is specific to the ULI II; however, with only slight modifications, the information also applies to a ULI I. The Signal Interface II is also designed for use with either a Macintosh or PC-compatible computer. The MPLI interface is designed for use only with PC-compatible computers.

In the Workshop Physics program, several sensors are needed to undertake the recommended activities. These sensors include a radiation sensor, an ultrasonic motion sensor, a force sensor, a temperature sensor, a voltage sensor, a magnetic field sensor, and a rotational motion sensor. This appendix contains a brief description of each of the sensors. Detailed information for the setup for each type of computer-based laboratory system and the use of the associated software are included in the system manuals.

THE UNIVERSAL LABORATORY INTERFACE (ULI)

The Universal Lab Interface (ULI) was designed to provide a way to connect a variety of sensors to any computer using a standard RS 232 or

* Portions of this Appendix are adapted with permission from apparatus and software manuals published by Vernier Software and PASCO scientific Company.

RS 422 serial port. Designed and manufactured by Transpacific Computer Company, the ULI is distributed by Vernier Software. The ULI is a small computer that monitors sensors attached to it and relays the sensor readings to a computer executing the proper software. The ULI provides a 12-bit Analog to Digital Converter (ADC), four 5-pin DIN (analog) connections, two 6-line telephone-style ports, and two audio-style jacks for digital inputs. The ULI has an open architecture to allow developers to design new sensors and software. A developer's guide is available from Vernier Software.

Fig. B.1. The Universal Lab Interface (front view)

The ULI is compatible with many sensors and several pieces of software that provide the functionality needed to perform experiments related to the Workshop Physics and Real Time Physics curricular materials. The software, developed at Tufts University for the ULI, runs on DOS or Windows based PC-compatibles and on Macintosh computers. The software is feature rich and allows significant flexibility; therefore, many experiments beyond those developed for Workshop Physics courses can be performed. Some features of the ULI software include real time data graphing, generation of tangent lines to graphs, integration, curve fitting, and the spectral analysis of data using a fast Fourier transform (FFT) routine.

PASCO SIGNAL INTERFACE II (SI2)

The Signal Interface II (SI2) was designed and is distributed by PASCO scientific. The SI2, similar to the ULI, provides a way to connect a variety of sensors to a Macintosh or Windows based PC-compatible computer. The SI2, however, uses a high-speed parallel connection to the computer via a Small Computer Signal Interface (SCSI) port. Use of the high-speed SCSI connection allows the SI2 to be used simultaneously for both data collection and experiment control. The SI2 provides four audio-style ports for digital input/output and three 8-pin DIN connectors for analog input/output. Input and output on the three analog channels are performed using a shared 12-bit ADC and DAC pair.

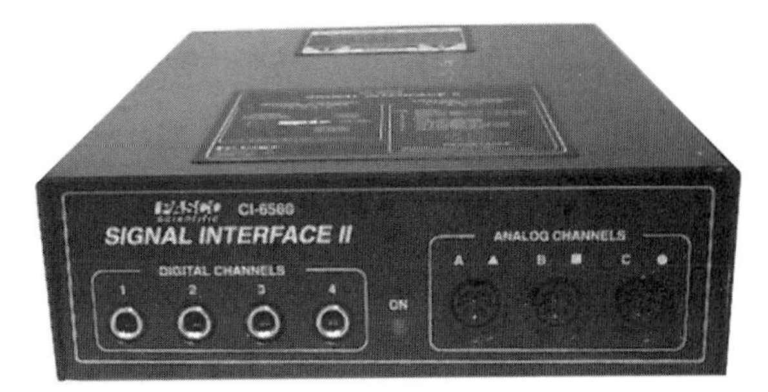

Fig. B.2. The Signal Interface II (front view)

The SI2 has a wide range of sensors and accessories that allows the attached computer to become a digital voltmeter, a triple-trace digital storage oscilloscope, an electronic stop watch, a digital frequency counter, a spectrum analyzer, a radiation monitor and much more. With the addition of the external Power Amplifier, the interface can be used as a DC power source and a function generator. The software for the SI2 is a completely integrated package called the Science Workshop which is functionally the same on both Macintosh and Windows based PC-compatible computers. The Science Workshop provides control and monitoring functions for all the capabilities of the SI2 and its associated sensors. Display and analysis options in the Science Workshop include plots, meters, histograms, data tables, curve fitting, integration, and FFTs.

MULTIPURPOSE LAB INTERFACE (MPLI)

The MultiPurpose Lab Interface (MPLI), distributed by Vernier Software, provides an interface that allows DOS® or Windows® based PC-compatible computers to be used as laboratory instruments. The MPLI interface can be used with any Windows® or DOS® based PC-compatible computer with at least one free Industry Standard Architecture (ISA) expansion slot. The main advantage of the MPLI is that it can be run on older DOS based machines (8086/8088/80286 processors). The MPLI comes with a custom ISA expansion board that must be inserted into the PC-compatible computer. This expansion board provides a high-speed parallel interface between the MPLI and the computer. The MPLI contains three 8-pin DIN (analog) ports that share a 12-bit ADC for monitoring sensor data. The MPLI also provides an internal 16-pin DIP socket for expansion circuits created by developers. A developer's tool kit is available from Vernier Software.

Fig. B.3. The MultiPurpose Lab Interface (front view)

A wide range of sensors is available for use with the MPLI interface. Many of the same sensors that are used with the ULI and the SI2 can also be used with the MPLI interface. Both DOS®and Windows® versions of the MPLI software are available. The Windows® version of the MPLI software supports all of the sensors and includes data analysis tools such as tangent lines, integration, a three channel storage oscilloscope, FFTs, histograms, spreadsheet-like calculated columns, and curve fitting. The DOS® version of the MPLI software supports all sensors except the motion sensor. To use a motion sensor with the DOS® based MPLI, a special software package called Motion Plotter must be used. The Motion Plotter software, when accompanied by an MPLI and the proper sensors, can graph distance, velocity, acceleration, and force. The DOS® based MPLI software has a three-channel storage oscilloscope, graphing capabilities, and real-time Fourier analysis.

SENSORS FOR THE COMPUTER-BASED LABORATORY SYSTEMS

The Motion Sensor

The Ultrasonic Motion Detector from Vernier Software is an analog sensor that connects to the ULI or MPLI computer interfaces. The ULI and motion detector are supported by the "Motion" software. The MPLI Windows based software supports the use of a motion sensor. To use a motion sensor with the MPLI and DOS® requires the Motion Plotter software. The PASCO scientific motion sensor is a digital sensor intended for use with the SI2. The SI2 and Science Workshop software support the use of two motion sensors simultaneously to record positions, velocities, and accelerations of both objects in a collision experiment.

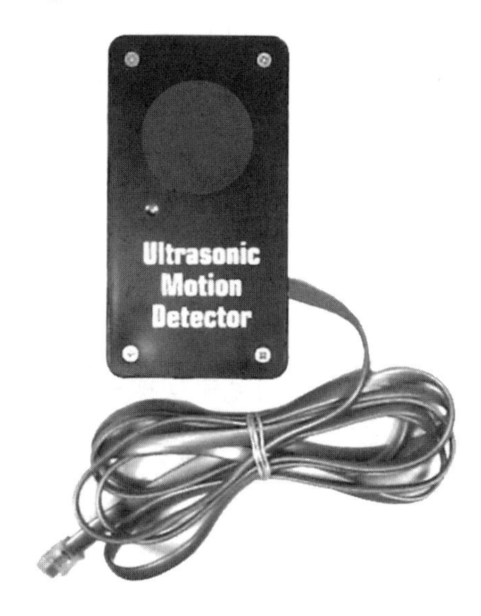

Fig. B.4a. Ultrasonic Motion Detector from Vernier Software is designed for use with the ULI (U-MD) and MPLI (MD-M).

Fig. B.4b. Motion Sensor from PASCO scientific is designed for use with the SI2 (CI-6529).

Note: The codes in parenthesis are the appropriate Vernier Software or PASCO scientific part codes for the sensors.

A motion sensor uses ultrasonic sound pulses

to measure the distance to the closest object located in front of the sensor (i.e., along its "line of sight"). The distance measured is always the distance to the object that is closest to the sensor and within the sensor's field of view. (The field of view is a 16 degree cone along the sensor's line of sight.) The range of distances that can be measured is from 0.4 m to 6 m for the Vernier sensor and from 0.4 m to 8.9 m for the PASCO sensor. If the location and orientation of the sensor are known, the distance measured is directly related to the position of the object. This subtle relationship between distance and position can be simplified by defining a coordinate system with the sensor at the origin and a position axis along the sensor's line of sight. In this coordinate system the distance measured by the sensor is equivalent to the one-dimensional position of the object.

To determine distances, the motion sensor emits pulses of ultrasound and receives reflections of these pulses after they bounce off objects within the sensors field of view. The frequency of this ultrasound is about 50 KHz. Since the speed of ultrasound in room temperature air is known, the computer motion software can calculate the distance of an object by timing how long the pulse takes to reflect off the object and return to the sensor. This is similar to how a bat "sees" and it is also how a Polaroid® auto-focus camera determines the distance to an object in order to focus properly.

The software packages for use with the motion sensors use distance measurements at evenly spaced times to calculate, using standard difference equations, the average values of velocity, and acceleration over each time period. Values of position, velocity, acceleration, and force can each be displayed in a variety of graphical formats as functions of time or as functions of each other. They may also be displayed in raw format as a data table. The motion sensor and its software form a useful tool for the analysis of complicated data from one-dimensional motions and forces.

The Force Sensor

There are two distinct types of force sensors recommended for use in the Workshop Physics activities. The Force Probe from Vernier Software is a Hall Effect based force sensor and is designed for use with the ULI. Alternatively there are a pair of strain gauge based force sensors from PASCO scientific. These strain gauge based force sensors are called the Student Force Sensor and the Force Sensor and are designed for use with the SI2. Versions of the Student Force Sensor and the Force Sensor are also available, from either PASCO scientific or Vernier Software, for use with the MPLI or ULI interfaces.

The Hall Effect Force Probe

A Hall Effect force sensor, like a motion sensor, actually measures distance. If you examine the sensor, you will see a cylindrical-shaped magnet attached to a strip of brass. The strip of brass acts as a leaf spring. The magnetic field surrounding the magnet is sensed by a Hall Effect transducer mounted at the end the force sensor's handle. For small flexures, the brass leaf spring acts like a Hooke's law spring, and therefore the position of the magnet is linearly related to the applied force. As the brass sheet that holds the permanent magnet flexes under a push or a pull, the magnetic field strength at the Hall Effect transducer changes. The transducer produces a current that varies linearly with the magnetic field strength. The current in the transducer is sensed electronically by the ULI and sent to the computer. The "Motion" or "Data Logger" computer software uses calibration information, and the linear nature of the relationships between the applied

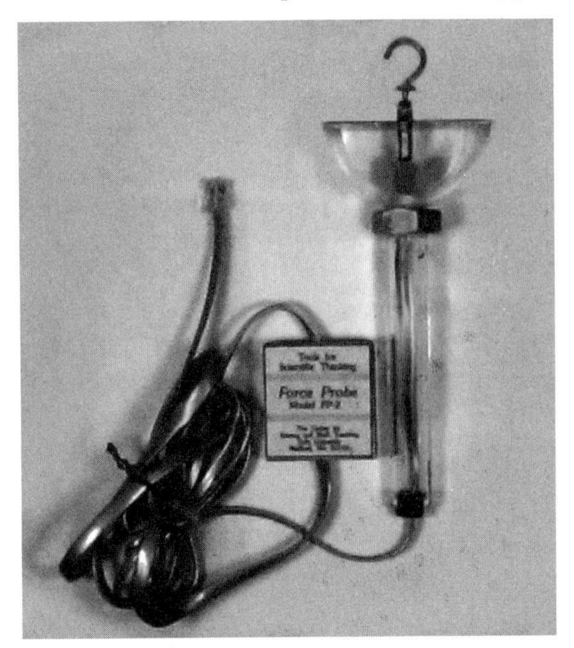

Fig. B.5a. Hall Effect Force Probe from Vernier Software for use with the ULI (U-FP).

force, the position of the magnet, and the Hall Effect transducer current, to calculate the applied force.

The sensitivity of the Hall Effect force sensor can be adjusted by moving the head of the sensor (which includes the magnet) relative to the handle. Moving the magnet closer to the Hall Effect transducer (built into the end of the handle) makes the force sensor more sensitive but limits the range of forces that can be measured. Moving the magnet further out makes the force sensor less sensitive and allows it to be used with a wider range of forces.

The Strain Gauge Force Sensors

The Student Force Sensor and the Force Sensor from PASCO scientific are both based on the measurement of strain in a metal beam. In the Student Force sensor, a pair of strain gauges are mounted on a single cantilevered beam that flexes under applied forces. As the beam flexes, the strain gauges are stretched or compressed, depending on the direction of the applied force. Stretching (pushing on the beam) or compressing (pulling on the beam) the strain gauge increases or decreases respectively the electrical resistance of the strain gauge. The change in resistance is used to generate an output voltage for the sensor

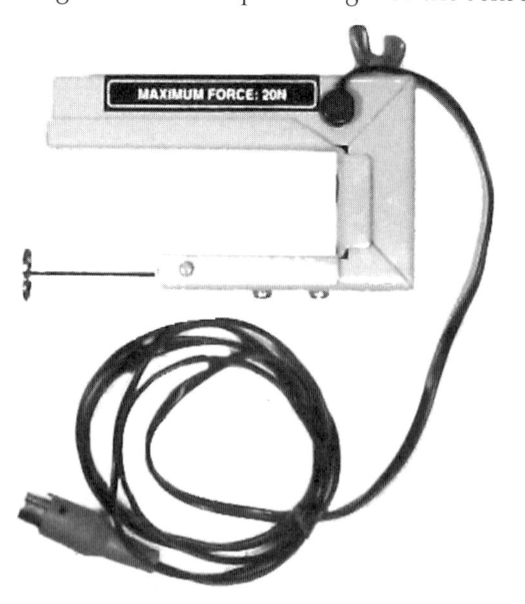

Fig. B.5b. Student Force Sensor from PASCO scientificis designed for use with the SI2 (CI-6519) and from Vernier Software for use with the ULI and MPLI (SFS-DIN).

Fig. B.5c. Force Sensor from PASCO scientific is designed for use with the SI2 (CI-6537) or the ULI and MPLI (: CI-6618 or Vernier Software: FS-DIN).

that is linearly related to the applied force. Therefore, the maximum voltage is output when the maximum positive force is applied to the sensor (a push of 20 N), and the voltage is a minimum when the maximum negative force is applied to the sensor (a pull of 20 N).

The Force Sensor is designed to have adjustable sensitivity, to be more accurate over a wider range of forces, and to be less susceptible to resonance effects during collision experiments than the Student Force Sensor. The Force Sensor is based on a standard "binocular beam," that looks like a metal plate with the front view of a pair of binoculars cut out of it. There are four strain gauges mounted on the beam, two on each side at the points where the cutout "binoculars" are closest to the edge. When an applied force creates a lateral strain in the beam, the strain gauges generate an output voltage that is proportional to the applied force. The Force Sensor can be adjusted, using a range select switch, to have a maximum range of –10 N (pull) to +10 N (push) or –50 N to +50 N. A tare button is also provided to allow the sensor to be zeroed manually.

The Science Workshop software for the SI2 and the DOS or Windows based MPLI Software have built in capabilities for the use of force sensors. With a ULI, the Motion or Data Logger software can be used. It is also common to use a force sensor in combination with a motion sensor to perform experiments to verify Newton's laws of motion. All of the above software packages are capable of these experiments with the exception of the DOS® based MPLI software. To perform these experiments with a DOS® based MPLI system, the Motion Plotter software is required.

The Radiation Sensor

The type of radiation sensor recommended for use in the Workshop Physics program consists of a Geiger counter and associated electronics which provide a signal output that can be attached to a digital input on a computer-based laboratory system. Suitable radiation sensors with computer output ports and cables are the Radiation Monitor distributed by Vernier Software and the Radalert and the LabNet Geiger-Müller Interface distributed by PASCO scientific. The Radiation Monitor and Radalert are hand-held units that can be used with or without a computer and interface system. The LabNet Geiger-Müller Interface requires the SI2 and Science Workshop or other compatible interface/software combination.

Fig. B.6a. The Radiation Monitor from Vernier Software.

Figure B.6b: The LabNet Geiger-Müller Interface from PASCO scientific is designed for use with the SI2 (SE-7997).

Fig. B.6c. The Radalert from PASCO scientific is designed for use alone, with the SI2, with the ULI and with the MPLI (SE-7996).

A radiation sensor is used to detect gamma particles or beta particles emitted from the nuclei of atoms in a radioactive source. In the study of physics, it is useful to count the number of particles that are detected to learn about such topics as counting statistics, natural radioactivity, and the half-lives of some relatively short-lived radioactive sources. All of the radiation sensors mentioned above detect particles in approximately the same manner. Each uses a stainless steel chamber, called a Geiger tube, that has been emptied of air and filled with a noble gas (usually Neon or Argon). When a beta or gamma particle, the result of a radioactive decay, passes through the Geiger tube, electrons are ripped away from the noble gas atoms. These electrons are attracted to a high voltage anode contained in the tube. As the electrons move to the anode, they knock more electrons free from the gas atoms, creating a cascade effect. The arrival of electrons at the anode creates a current. This current drives an electronic circuit that sends a signal to the computer. The computer interprets the signal that is received as an "event." Several other electrical and chemical processes allow the noble gas ions to acquire new electrons and return to their neutralized ground state.

Counting software designed to work with a computer interface allows you to count the discrete events detected by a sensor. For the ULI, the counting software is the "Event Counter" program. The Science Workshop program for the SI2 and the MPLI software for the MPLI both contain counting capabilities. Events that can be counted are not limited to the presence of particles emitted from a radioactive source but depend on the sensor that is used and include pulses of light or sound bleeps. The software programs allow the number of counts in an interval of time to be displayed graphically or in a data table. The software can calculate averages and other statistics for repeated trials. Data can be displayed in the form of a graph of counts/interval vs. time, or as a histogram of frequency vs. counts/interval or in both formats simultaneously.

The Voltage Sensor

The analog ports on all of the interfaces accept analog voltages as input. These voltages are converted to a binary number using a 12 bit ADC (a ULI I has a 10 bit ADC). The Science Laboratory software with the SI2 and the MPLI software with the MPLI interface allow voltages between ±10 V to be measured. These software packages also allow the SI2 and MPLI to be used as three-channel digital storage oscilloscopes. The ULI with the Data Logger software is able to measure voltages between 0 and 5.12 V. Exceeding these maximum voltages may cause damage to the interfaces.

Fig. B.7a. Voltage Measurement Leads from Vernier Software is designed for use with the ULI, MPLI and SI2 (TL-DIN).

Fig. B.7b. Voltage Sensor from PASCO scientific is designed for use with the SI2 and MPLI (CI-6503).

The Temperature Sensor

There are a wide range of temperature sensors available from both Vernier Software and PASCO scientific, only two of which are shown in Figure B.8. These temperature sensors use a precision temperature-sensitive integrated circuit whose output is linearly related to temperature. The sensors are covered in a Teflon® heat shrink tubing to make them resistant to chemical solutions including oxidizing agents and organic solvents. The Vernier Standard Temperature Probe comes with a small black box that contains signal-conditioning circuitry and has a temperature range of –50 °C to +150 °C. The PASCO Temperature Sensor connects directly to the SI2 and has a temperature range of –5 °C to +105 °C. The probes should not be operated outside these ranges or damage may occur.

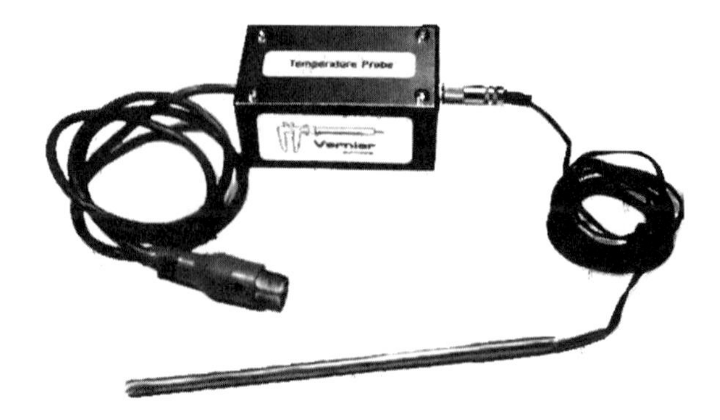

Fig. B8a. Standard Temperature Probe from Vernier Software is designed for use with the ULI and MPLI (TPA-DIN).

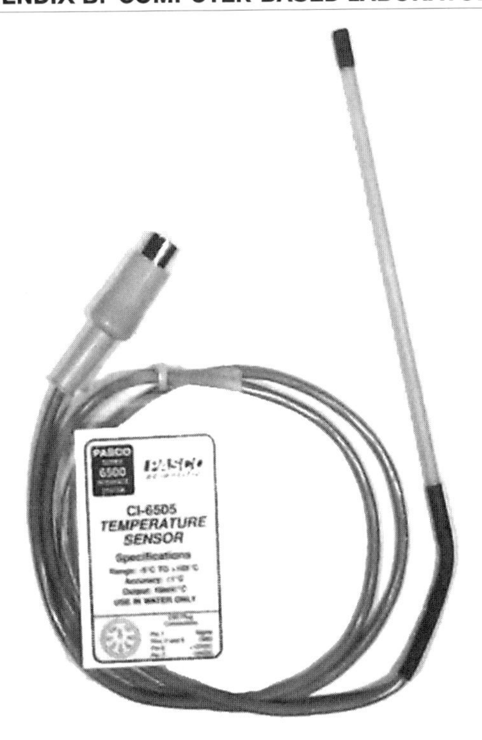

Fig. B.8b. Temperature Sensor from PASCO scientific for use with the SI2 (CI-6505A).

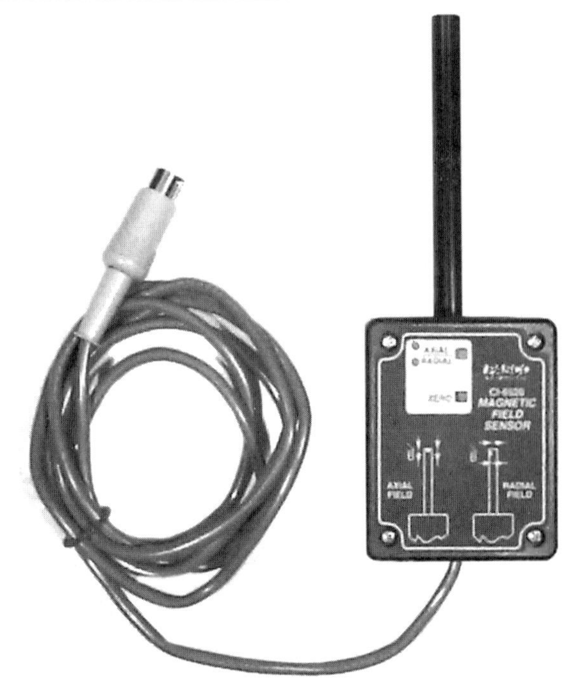

Fig. B.9a. Magnetic Sensor from PASCO scientific is designed for use with the SI2 (CI-6520).

The Magnetic Field Sensor

Both the Magnetic Field sensors shown in Figure B.9 are based upon Hall Effect transducers. A Hall Effect transducer produces a current that varies linearly with the magnetic field strength at the location of the transducer. The current in the transducer is sensed electronically by the computer interface and is sent to the computer.

The Vernier Magnetic Field Sensor has a single Hall Effect transducer. This single transducer measures the magnetic field strength parallel to the hand-held wand of the sensor. The sensitivity of the sensor can be selected to measure magnetic field strengths in two ranges, either ±3.2 Gauss or ±64 Gauss. When using the ±3.2 Gauss range, this sensor is sensitive enough to measure the strength of the Earth's magnetic field.

The PASCO Magnetic Sensor contains two Hall Effect transducers installed at right angles to each other. Therefore, the magnetic field strength at the location of the transducers can be measured in two directions. However, the two perpendicular field strengths can not be measured simultaneously. This sensor measures magnetic field strength in the range of 10 Gauss to 2000 Gauss with a resolution of approximately 10 Gauss. Therefore, the Magnetic Sensor can not be used to measure the strength of the Earth's magnetic field.

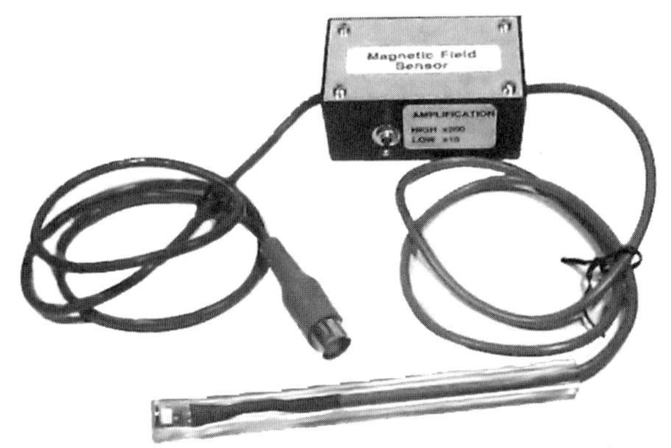

Fig. B.9b. Magnetic Field Sensor from Vernier Software for use with the ULI and MPLI (MG-DIN).

The Rotary Motion Sensor

The PASCO and Vernier Rotary Motion Sensors use optical encoders to measure the angular displacement of a shaft. An optical encoder consists of a transparent disk with a known number of opaque lines etched radially at evenly spaced intervals. The spacing of these lines dictates the resolution of the optical encoder. There are two photogates, each of which creates a closed circuit when its light beam passes through a transparent section of the disk. Conversely, the circuit is broken when the light beam is blocked by one of the opaque lines. As the probe is rotated, the output of each of the photogates is a square wave, because its light beam is repeatedly blocked and unblocked. The two photogates are slightly offset, so that as the disk rotates clockwise, photogate 1 is blocked before photogate 2. If the disk rotates counter-clockwise, then photogate 2 will be blocked before photogate 1. Therefore, the direction of rotation can be determined by the phase relationship of the square waves from the two photogates.

Fig. B.10. PASCO Rotary Motion Sensor (CI-6538). The CI-6625 is available for use with the ULI.

The software that is used with the rotary motion sensors counts the number of times a photogate is blocked to determine the amount the probe has been rotated. The software allows the probe to be zeroed at any point, allowing a direct correspondence between the angular displacement measured by the probe and the actual angular position. The direction of rotation is determined by the software, which monitors the phase relationship between the square waves from the photogates. The software also calculates angular velocity and angular acceleration by using difference equations. For the version of the PASCO Rotary Motion Sensor used with the ULI, the "Rotary Motion" software performs these functions as well as a variety of data analysis functions similar to those in the "Motion" software. A model of the PASCO Rotary Motion Sensor is also available for use in conjunction with the Science Workshop software and SI2 interface.

Additional and up-to-date information about the Computer-Based Laboratory systems, software, or sensors described in this appendix can be obtained from:

PASCO scientific

P.O. Box 619001 • 10101 Foothills Blvd.
Roseville, California 95678-9011
Phone: (916) 786-3800
Fax: (916) 786-8905
E-mail: sales@pasco.com

Vernier Software

8565 S.W. Beaverton-Hillsdale Hwy.
Portland, OR 97225-2429
Phone: (503) 297-5317
Fax: (503) 297-1760
E-mail: dvernier@vernier.com

APPENDIX C
STATISTICAL MEASURES OF UNCERTAINTY

There are three kinds of lies: lies, damned lies, and statistics.
Benjamin Disraeli (according to Mark Twain)

UNCERTAINTY IS A FACT OF MEASUREMENT

There are a few quantities that can be determined in principle with complete certainty. For example, if discrete things are counted, the degree of precision can be infinite. Everyone can agree that there are 4 dogs and 9 cats at the animal shelter. Also, certain mathematical entities such as π, which represents the ratio of the circumference to the diameter of a circle, can be known to any desired precision. This is because a perfect circle is an abstract mathematical entity. Although π is an irrational number, it can be calculated to as many decimal places as desired e.g., 3.1415927 . . . and so on.

Many of the measurements you will make in the study of the physical world are imprecise. The length of an object, the temperature of a cup of liquid, the time of fall of a ball. There are inevitable uncertainties associated with all of these types of measurements. There are no measuring instruments capable of infinite precision when the quantities being measured can vary continuously. Thus, in the physical sciences it is often as important to know how certain a given numerical result is as it is to know the result itself. After all, much of science consists of comparing a model world of mathematical theory to the "real" world we try to learn about in the laboratory, the field, or heavens. To make sense of measurements, we must have a good idea of just how closely we should expect the two worlds to compare. As a general rule, almost as much work goes into an honest estimate of the uncertainty of a result as went into deriving the result itself. Ideally, every number you write down when making a measurement or performing a calculation based on measurements should be evaluated as to how certain it is.

SYSTEMATIC ERRORS VS. STATISTICAL UNCERTAINTIES

There are two kinds of "uncertainties" associated with measurements: (1) *Systematic errors,* which increase or decrease all measurements of a quantity in the same sense (either all measurements will tend to be too large, or tend to be too small), and (2) *statistical uncertainties,* which are completely random.

SYSTEMATIC ERRORS

Systematic errors usually arise due to consistent inaccuracies in measuring equipment. For example, a spring scale may always read too large a force. Systematic errors can also arise from human errors such as an investigator consistently misinterpreting a number marked on a scale. These errors can also result from the neglect of special conditions. For example, in measuring length with a steel ruler, one should take into account the fact that the ruler expands and contracts very slightly with temperature, and is strictly accurate at only one temperature.

Systematic errors are very hard to deal with, but once determined, such errors can be removed from the reported results. For example, one could in principle take a good quality ruler to the National Bureau of Standards and compare it to a very accurate standard meter stick there and find exactly how much larger or smaller a length it tends to read at a standard temperature. One could then multiply the previous results by the appropriate factor to correct the results.

Sometimes, if you cannot pinpoint the exact effect of a systematic error in ordinary laboratory practice, you can make a reasonable estimate (read "guess") of how important your systematic errors might be. Then you can correct your data to take this error into account on an approximate basis.

STATISTICAL UNCERTAINTIES

Statistical uncertainties arise from a series of small, unknown, and uncontrollable events. As a result, it is impossible for repeated measurements of the same quantity to yield precisely the same value. (If you read a scale or a meter as carefully as you can and are not biased by earlier readings, you will notice small differences in each of your readings.) Often, the best value of a series of readings is given by the average of the readings. Statistical uncertainties are often easier to report than systematic errors, because there are definite mathematical rules for estimating their size.

ESTIMATING UNCERTAINTIES BASED ON MEASUREMENTS

Overview

Intuitively, we would expect the uncertainty in a measurement to be about the size of the range in readings from the lowest to the highest. In general, repeated measurements of the same phenomenon will not yield the same value. Thus, we usually attribute these differences to random unknown events that change the exact conditions of the measurement slightly from reading to reading. These random events lead to statistical uncertainties. Let's say, for example, that we have a series of measurements of a length: 12.2, 12.2, 12.3, 12.0, 12.1 cm. The overall range in our data would be 12.0 to 12.3 cm, and we would expect our measurements to be no more uncertain than 0.3 cm.

Statisticians have developed a theory of uncertainty that can put our intuition on a firm footing. There are two mathematical quantities that are commonly used to describe statistical uncertainties: the *standard deviation* (or σ_{sd} for short), and the *standard deviation of the mean* (or SDM for short). The SDM is sometimes called "standard error"—a term we avoid using since the term error in this sense is a misnomer.

When taking measurements in physics, we can often assume that variations in measurements are random, or "normally" distributed—just as likely to be too high as too low. If this is the case, then both the *standard deviation* and the *standard deviation of the mean* can be derived from a set of repeated measurements. The procedure for deriving these terms will be discussed later in this appendix.

Mistakes Are Not Uncertainties

There are, of course, true errors—that is, mistakes. Among these are mistakes in calculations and errors in the use of equipment. Errors can never be completely avoided. It is your responsibility to check and double check calculations and to be sure that equipment is used properly so that errors are minimized. However, statisticians and experimental scientists often use the term *statistical error* or *error* to mean uncertainty rather than mistake or blunder.

The Standard Deviation

The standard measure of the uncertainties associated with a series of repeated measurements is known as the *Standard Deviation*. The Standard Deviation answers the question "If, after having made a series of measurement and averaging the results, I now make one more measurement, how close is the new measurement likely to come to the previous average?" Here is how you would calculate it if you had a sample of 12 (or more generally N) measurements:

a. Sum all the measurements and divide by 12 to get the *average* or mean.

b. Now, subtract this *average* from each of the 12 measurements to obtain 12 "*residuals.*"

c. Square each of these 12 *residuals* and add them all up.

d. Divide this result by $(N - 1)$ (in this case 11) and take the square root.

Lo and behold, you have the standard deviation! We can write out the equation for calculating the Standard Deviation:

Let each of the N measurements be called x_i (where $i = 1$ to N) and let the average of the N values of x_i be μ. Then each residual $r_i = x_i - \mu$. Thus:

$$\mu = \frac{(x_1 + x_2 + x_3 + x_4 + \ldots + x_N)}{N} \quad \text{(C.1)}$$

$$r_i = \mu - x_i$$

$$\text{SDM} = \sigma_{sd} = \sqrt{\frac{(r_1^2 + r_2^2 + r_3^2 + r_4^2 + \ldots + r_N^2)}{(N-1)}}$$

$$= \sqrt{\frac{\sum r_i^2}{(N-1)}}$$

Here the symbol Σ means "sum the terms $i = 1$ to $i = N$."

The standard deviation represents the uncertainty in any one measurement. It is often instructive to view your measurements, x_i, in histogram form. To do this you need to divide the overall range of your measurements into a set of smaller ranges. The histogram is a plot of the number of measurements that fall within each small range vs. the average value of the small range. If you took 50 or 100 measurements, instead of 12, and plotted a histogram, you might get a graph that looked like the one shown below in Figure C.1.

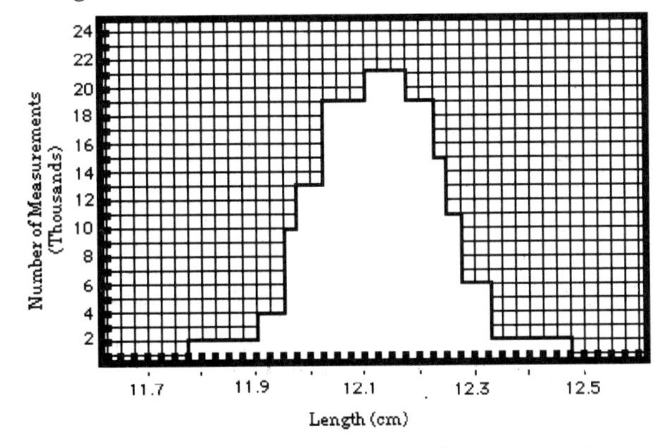

Fig. C.1. A histogram representing the variation in a set of measurements. The height of each bar is proportional to the number of measurements in each small range of values.

This graph is a *histogram*. It represents the number of times a particular value of a measurement, in this case a length, is made for each 0.25 cm interval of length measurement. For example, the histogram represents the fact that 10 of the 100 measurementsyielded lengths between 11.950 cm and 11.975 cm. Clearly, values near the middle of the overall range come up more frequently than do values at either end of the overall range. If we use the formulas discussed above on the data plotted here, we find that the average or mean lies close to the middle of the range. From this graph one could estimate a mean of 12.1 cm or so and a probable range of from 12.0 to 12.3 cm.

If, instead of 100 measurements, you took several thousand and divided the overall range into very small ranges, your graph could be drawn as a smooth curve, such as the one shown below. For this "bell-shaped" curve (also called a "normal distribution" or a "Gaussian") most points lie somewhere between the average value and one standard deviation away from the average value. Since one standard deviation (one 'sigma,' σ_{sd}) in this graph is about 0.1 cm, most points lie between 12.0 and 12.2 cm. In fact, it can be proved mathematically that for completely random uncertainties, about 68% of the points will lie within one standard deviation (one σ_{sd}) from the mean.

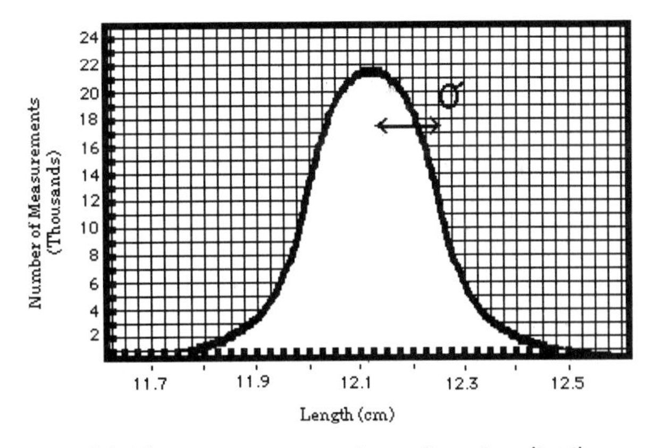

Fig. C.2. The curve representing a Gaussian distribution. This is often called the "bell-shaped" curve.

Standard Deviation of the Mean (alias Standard Error)

To get a good estimate of some quantity you need several measurements. You really want to know how uncertain the *average* of those measurements is, since it is the average that you will write down (as a best estimate). This uncertainty in the average is what is known as the *Standard Deviation of the Mean* or the *Standard Error.* Once the standard deviation has been calculated for a sample containing N measurements, it is very easy to calculate the standard deviation from the mean using the equation:

$$\mathrm{SDM} = \frac{\sigma_{sd}}{\sqrt{N}} \qquad \text{(C.2)}$$

It is this quantity that answers the question, "If I repeat the entire series of N measurements and get a second average, how close can I expect this second average to come to the first one?" One need not repeat the entire experiment multiple times to get the standard deviation from the mean. If the variation in data seem to be caused by a seriesof small random events, then equation C.2 can be used to find the standard deviation from the mean. It is this standard deviation from the mean that one usually reports as the final uncertainty. The abbreviation that is often used for the Standard Deviation of the Mean is SDM Sometimes the SDM is referred to as standard error. Since the SDM is actually a measure of uncertainty rather than of an error, in the sense of a mistake, we prefer not to use the term standard error.

Sample Calculation of Standard Deviation and SDM

Let us imagine that you made the following series of length measurements with a good centimeter ruler: 12.2, 12.2, 12.3, 12.0, 12.1 cm.

	Measure	Mean	Residual	Residual^2
1	12.2	– 12.16=	+0.04	0.0016
2	12.2	– 12.16=	+0.04	0.0016
3	12.3	– 12.16=	+0.14	0.0196
4	12.0	– 12.16=	+0.16	0.0256
5	12.1	– 12.16=	–0.06	0.0036
	Sum: 60.8			Sum: 0.0520

Average:

$$60.8/5 = 12.16 \text{ cm}$$

Sum of residuals squared:

$$0.0520$$

Sum of residuals squared divided by $(N–1)$:

$$0.0520/4 = 0.0130$$

Standard deviation:

$$\sigma_{sd} = \sqrt{0.0130} = 0.114 \text{ cm}$$

Standard deviation of the mean:

$$\mathrm{SDM} = 0.114/\sqrt{5} = 0.051 \text{ cm}$$

Reported result:

$$L = 12.16 \pm 0.05 \text{ (SDM) cm}$$

Here the SDM indicates that you are quoting the standard deviation of the mean as your measure of the uncertainty in your best estimate. This represents a 68% confidence level.

THE 95% CONFIDENCE LEVEL

Often when you are asked to report data based on measurements, we would like to have you report the mean or average along with a 95% confidence interval with a "± " (plus or minus) sign in front of it. We will use the notation S(95) for this quantity. S(95) is shown on the idealized histogram in Figure C.3. The area under the dark gray part of the probability curve is 95% of the total area under the curve.

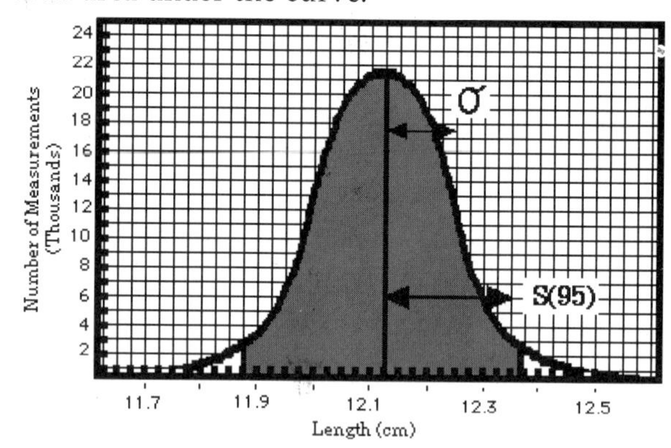

Fig. C.3. A Gaussian distribution curve showing the 95% confidence interval.

If you were to take an infinite number of data points, S(95) would simply be twice the standard deviation of the mean (SDM). The Excel 5.0 spreadsheet we are using in the Workshop Physics course comes with an add–in Macro function in its Analysis ToolPak that allows us to calculate S(95) based on a smaller number of measurements. In general, the S(95) obtained from the Excel Toolpak and 2σ are not quite the same, but they tend to be similar in magnitude.

SIGNIFICANT FIGURES

Whenever a value is obtained from a measurement or as part of a calculation involving measured values, it is very useful to establish rules for expressing the value with a "correct" number of significant figures.

It is difficult to learn to use and interpret significant figures because scientists tend to use slightly different definitions and rules for determining them. For example, if the time of fall of a ball is measured repeatedly and the best estimate (average) is calculated as $t = 0.46236748$ seconds and the standard deviation is calculated as 0.01524267, one experimenter might report $t = 0.46 \pm 0.02$ seconds and another might retain one more significant figure in both the best estimate and uncertainty and report $t = 0.462 \pm 0.015$ seconds. However, they will both agree that 0.5 seconds is too few significant figures to report and that 0.4623 is too many.

Note: In Workshop Physics activities and assignments, you should learn the rules well enough to report significant figures in an acceptable range in both handwritten assignments and in spreadsheet cells.

Estimating the Best Value of a Measured Quantity

Generally, one arrives at a best estimate of a quantity of interest by making a series of measurements and averaging the results. Statisticians will tell you that, as a rule of thumb, you should make at least eight measurements of a quantity before you can "do statistics" to properly estimate averages and uncertainties! This is rarely convenient, and often impossible to do in the physics laboratory, but, whenever you can, you should make at least three measurements of any given quantity.

With a minimum of three measurements, you stand some chance of spotting problems or out-and-out blunders. If one of the measurements is radically different from the other two, you can check your results by taking more data. Do not, however, exclude a measurement from the final average unless you have good evidence that the measurement was in error.

Reporting Results of Measurements with Uncertainties

We usually report a result in the form: best estimate ± uncertainty. Generally, the uncertainty is given to one (or possibly two) significant figure(s). That amounts to one (or two) non-zero number(s) after the decimal point. *The least significant figure reported for the best estimate of the result always should be of the same order of magnitude as the uncertainty.* For example, the result of an experiment to measure the distance between two trees might sensibly be reported as 12.8 ± 0.2 m. The best estimate is given to just one decimal place, as is the uncertainty.

A much more precise experiment might yield the result 12.835 ± 0.004 m. The uncertainty is given to one significant figure, but it is in the third place after the decimal point, and so it is reasonable that the best estimate is carried out to three decimal places. It would, however, make no sense to report a result of 12.8 ± 0.2345678 for the first, low-precision experiment, nor would it make sense to state the result as 12.843567239 ± 0.234567823!

Implied Uncertainty and Significant Figures

Implied Uncertainty

If an uncertainty is not listed with a number that has been obtained from a measurement or from a calculation based on a measurement, there is an implied uncertainty associated with the number. For example, to a scientist the number 7 and the number 7.0 are not the same.

a. In a case in which the number 7 has been reported, there is only one significant figure. The scientist would have reported the number 6, if the result was actually closer to 6 than to 7. Or, the number 8 would be reported if the number were closer to 8 than to 7. The reporter is implying that the uncertainty associated with the number 7 is at least ±0.5. It may be more!

b. In a case in which the number 7.0 has been reported, there are now two significant figures. The scientist would have reported the number 6.9 if the result was actually closer to 6.9 than to 7.0. Or the number 7.1 would be reported if the number were closer to 7.1 than to 7.0. The reporter is implying that the uncertainty associated with the number 7.0 is at least ±0.05. It may be more!

Significant Figures
If we assume the reporter of a measured value has been scientific, then the number of significant figures is the same as the number of digits displayed. There are some exceptions to this rule:

a. Leading zeros that are located to the left of a decimal point don't count unless all the digits are zeros.

b. If all the digits in a number are zeros, then they are all significant.

c. If the last number of numbers of a string of whole numbers are zeros, the number of significant figures is often ambiguous unless it is re-expressed in scientific notation. (See the example in Table C1.)

Applications of the Definitions
Table C1 illustrates the rules for finding the implied uncertainties and the correct number of significant figures with some examples.

Handling of Significant Figures in Calculations
Properly, the correct number of significant figures to which a result should be quoted is obtained via error analysis. (See Appendix F for more details.) However, error analysis takes time, and frequently in actual laboratory practice it is postponed. In such a case, one should retain enough significant figures that round-off error is no danger, but not so many as to constitute a burden. The following is an example.

The reporter drops too many significant figures:

WRONG: $0.77 \times 1.46 = 1.1$

In this case the numbers 0.77 and 1.46 are known to be precise to about 0.5%, whereas the incorrectly reported result of 1.1 is only precise to about 5%. In this extreme case, the precision of the result is reduced by almost a factor of ten, due to round-off error.

The reporter keeps too many significant figures:

WRONG: $0.77 \times 1.46 = 1.1242$

The extra digits, which are not really significant, are just a burden, and in addition they carry the incorrect implication of a result of absurd accuracy.

The reporter keeps about the right number of significant figures:

OKAY: $0.77 \times 1.46 = 1.12$

Less good, but acceptable:

$0.77 \times 1.46 = 1.124$

Table C1. Examples of the Rules for Finding Implied Uncertainties and the Correct Number of Significant Figures.

Number based on a measure or estimate	Significant figures	Minimum implied uncertainty	Comments
7	1	±0.5	Implies the number is between 6.5 and 7.5
7.0	2	±0.05	Implies the number is between 6.95 and 7.95
7.00	3	±0.005	Implies the number is between 6.995 and 7.995
0	1	±0.5	Implies the number is between −0.5 and +0.5. Even zero can have significant figures!
0.00	3	±0.005	Implies the number is between −0.005 and +0.005
0.128	3	±0.0005	Implies the number is between 0.1275 and 0.1285. The leading zero is unnecessary, but it does help the reader notice the decimal point.
2.324	4	±0.0005	Implies the number is between 2.3235 and 2.3245
2.324×10^5	4	$\pm 0.0005 \times 10^5$	Implies the number is between 2.3235×10^5 and 2.3245×10^5
230	2 or 3	?	Ambiguous. The zero may be significant or it may be present only to show the location of the decimal point.
2.30×10^3	3	$\pm 0.005 \times 10^3$	There is no ambiguity if scientific notation is used. It is implied that the number is between 2.295×10^3 and 2.305×10^3.
2.3×10^2	2	$\pm 0.05 \times 10^3$	Again scientific notation clears up the ambiguity. It is implied that the number is between 2.25×10^3 and 2.35×10^3.

Rule 1: In multiplication or division it is often acceptable to keep the same number of significant figures in the product or quotient as are in the least precise factor.

Examples of Rule 1 (Multiplication and Division):

$$3.6 \times 25.7 = 92.52 = 92$$

$$\frac{4.6}{757} = .006077 = .0061$$

Rule 2: In addition or subtraction it is often acceptable to keep the same number of decimal places in the sum or different as are in the number with the least number of decimal places.

Examples of Rule 2 (Sum and Difference):

$$\begin{array}{r} 37.6 \\ \underline{-\ 2.45} \\ 35.2 \end{array} \qquad \begin{array}{r} 6953 \\ \underline{-\ \ 42.7} \\ 6911 \end{array} \qquad \begin{array}{r} 22.7 \\ 19.51 \\ \underline{+\ \ .732} \\ 42.9 \end{array} \qquad \begin{array}{r} 1.3378 \\ 15.43 \\ \underline{+1.821} \\ 18.58 \end{array}$$

BIBLIOGRAPHY

Two of the more informative books on measurement uncertainties are:

Baird, D.C., *Experimentation: An Introduction to Measurement Theory and Experiment Design* (Prentice-Hall, Englewood Cliffs, NJ, 1988)

Taylor, J.R., *An Introduction to Error Analysis: The Study of Uncertainties in Physical Measurements* (University Science Books, Mill Valley, CA, 1982)

APPENDIX D
GRAPHING DATA WITH UNCERTAINTIES–ERROR BARS AND EYEBALLS

One should accept the hypothesis that explains all the data with the minimum number of assumptions.

Paraphase of a principle known as "Occam's Razor"
William of Occam

OVERVIEW

There are many physics experiments in which data can be used to tell us something about the relationship between two quantities. For example, when a body falls down an inclined plane, how does its velocity depend on its position? How does its position depend on time?

One of the best ways to begin an exploration of the mathematical relationship between two variables is to represent the data on a graph. If the data points seem to lie along a familiar curve, we can often guess what the nature of the mathematical relationship is. Although there are formal techniques for finding an equation whose graph fits the data, graphing data and drawing a curve by eye is a good beginning. We will call this method of curve fitting the *eyeball method.*

Indicating Uncertainties on a Graph

It is impossible to record experimental data without uncertainties associated with each of the data points. For example, when a body falls down an inclined plane, repeated measurements under similar conditions would reveal some uncertainty in both the exact position of the body at a given time and its average velocity at that time.

It turns out that the nature of the curve that we might draw through the data points will depend on how large the uncertainties associated with each data point are. Fortunately, a standard method for indicating these uncertainties on a graph has been devised. In a graph of the relationship between two experimentally determined variables, one usually indicates the magnitude of the uncertainties in these variables by *error flags* or *error bars.* (Please remember that the word "error" here is really a misnomer. Although "uncertainty" is what is meant, "error bar" is the term most commonly used.)

A typical graph of experimental data, with error bars indicating the magnitude of the uncertainty in each of the variables, is shown below.

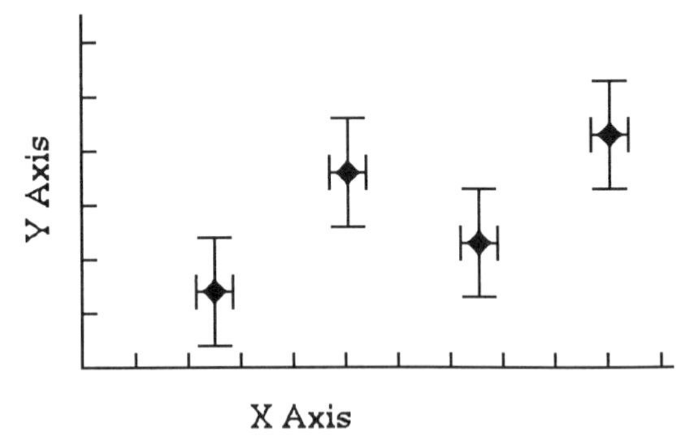

Fig. D.1. A graph of data that includes error bars that represent the uncertainty in both the x and y values.

Often, one of the physical variables will have an uncertainty so small that it can, for all practical purposes, be neglected. For example, in a typical motion experiment using a motion detector, the distance that an object has traveled in a certain time will probably be more uncertain than the time. In such an instance a simplified graph with error bars like those shown in the following figure could be drawn.

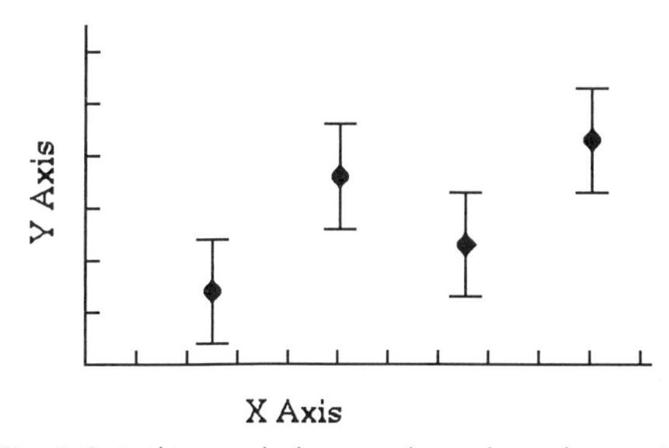

Fig. D.2. In this graph the error bars along the x axis are ignored because they are assumed to be much smaller than those along the y axis

HOW SOCIAL SCIENTISTS MIGHT FIT THE DATA

If we were social scientists, and the data in this graph represented popular attitudes about a given social issue as a function of time, we might be tempted to "connect the dots" as shown below:

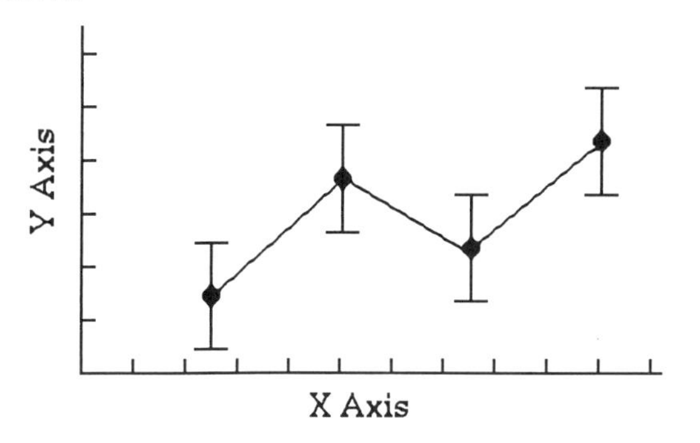

Fig. D.3. In this graph the person doing the curve fit has ignored uncertainty in the data.

Indeed, there is nothing that requires human behavior or attitudes to behave in a way that is continuous and differentiable! In fact, it is this difficulty of mathematically modeling human behavior that has made progress in the social sciences so problematic.

HOW PHYSICAL SCIENTISTS MIGHT FIT THE DATA

In the physical sciences, we would like to believe that the universe is smooth and continuous and simple. Therefore, a physical scientist would try to *draw the simplest possible curve that is compatible with the data and its associated uncertainties.*

Thus, if the uncertainties associated with our four points are large enough, a simple straight line such as the one shown in Fig. D.4 would represent the best fit.

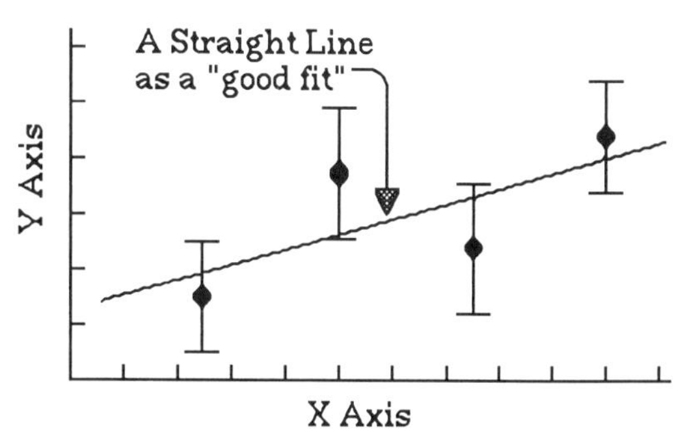

Fig. D.4. In this graph a straight line is the simplest "curve" that can pass through all the error bars.

On the other hand, if the uncertainties associated with the four sample data points are relatively small, a more complex curve that almost "connects the dots" is in order. This is shown below:

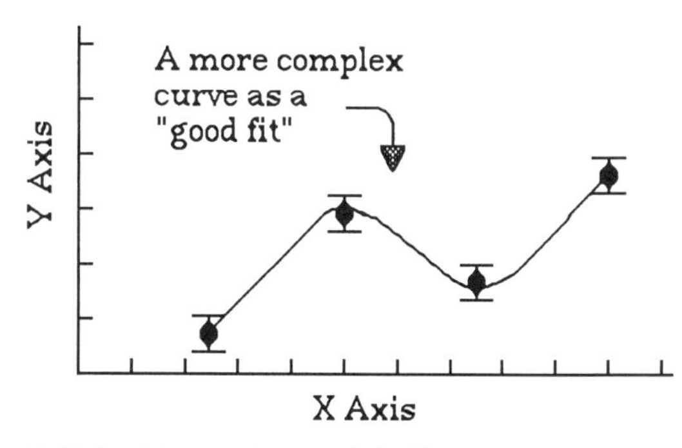

Fig. D.5. In this graph a straight line cannot pass through all the error bars.

Note that it is impossible to draw a straight line that passes through all the error bars on the graph in Fig. D.5.

SUMMARY

When data are represented graphically, any uncertainties in either of the quantities being measured should be represented on the graph as error bars or flags.

In using the eyeball method to fit the best curve to a set of data points with error bars, *the simplest curve that fits through most or all of the error bars should be drawn.*

This requirement that the simplest possible relation be used to fit the data is an example of a venerable principle in science called "Occam's Razor" or the principle of parsimony, named for William of Occam, an English scholastic philosopher of the fourteenth century. As usually stated, Occam's Razor says that one should accept the hypothesis that explains all the data with the minimum number of assumptions.

APPENDIX E
MATHEMATICAL MODELING TO FIT DATA

THE IDEA OF MODELING

One of the most powerful and elegant ways to represent simple relationships between two or more quantities in a physical system is by using mathematical equations. Sometimes, when examining a graph of data representing how one quantity changes with another, you'll find that the data points tend to lie along a straight line or a fairly smooth curve.

Fitting a Straight Line

The horizontal distance moved by a baseball after it leaves the pitcher's hand might appear to increase more or less linearly as shown in Figure E.1.

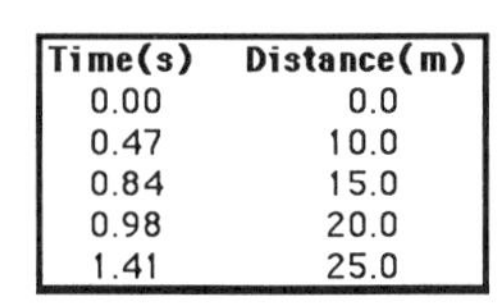

Time(s)	Distance(m)
0.00	0.0
0.47	10.0
0.84	15.0
0.98	20.0
1.41	25.0

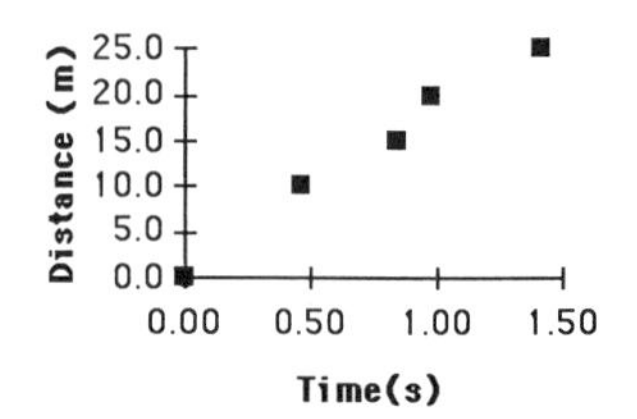

Fig. E.1.

In general, the equation for a straight line is represented by the equation

$$y = mx + b$$

where x represents an independent variable plotted on the horizontal axis and y represents a dependent variable plotted on the vertical axis. The m represents the slope of the graph and the b represents the y-intercept of the graph (i.e., the value of y when x is zero). Thus, we would expect that we could model the horizontal motion of the pitched baseball being described in the graph shown in Fig. E.1 by using the equation $Distance = m \times time + b$. In this case the y-intercept, b, is obviously zero. The data points don't seem to lie exactly along a line, just approximately. What is the best value for the slope, m? One way to estimate a reasonable value is to plot *Distance* vs. *Time* for a number of different values of m and see which line seems to pass closest to most of the data points. Various possible slopes are shown in the graph below.

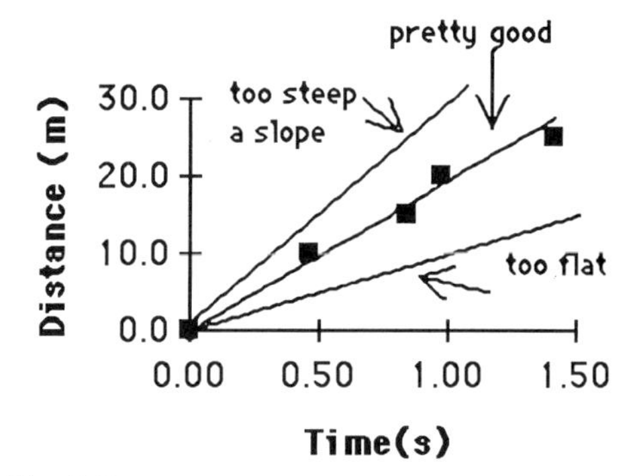

Fig. E.2.

The equation for the line on the graph that looks pretty good turns out to be

$$\text{Distance (m)} = (19.5 \text{ m/s}) \times \text{Time (s)}$$

Thus, the slope of a pretty good line is 19.5 meters per second and its y-intercept is 0.0 m. This process of looking at the shape of a graph and finding an idealized equation that more or less fits the data is called *mathematical modeling*. In our simple linear, or straight line, example, the equation that results is a very elegant shorthand for describing the horizontal progress of a particular pitched baseball as it travels. The purpose of mathematical modeling is to find equations and appropriate constants that seem to best describe a given physical system.

Modeling a "Curve"

Not all data points lie along a straight line. For example, let's consider some idealized data on the height of a miniature rocket as a function of time. These data are shown in Table E.1.

Table E.1. Model Rocket Data

Altitude (ft)	Time (s)	Time^2 (s^2)
0.0	0.0	0.0
32.0	1.0	1.0
128.0	2.0	4.0
288.0	3.0	9.0

A plot of altitude vs. time would yield a curve as shown in Figure E.3.

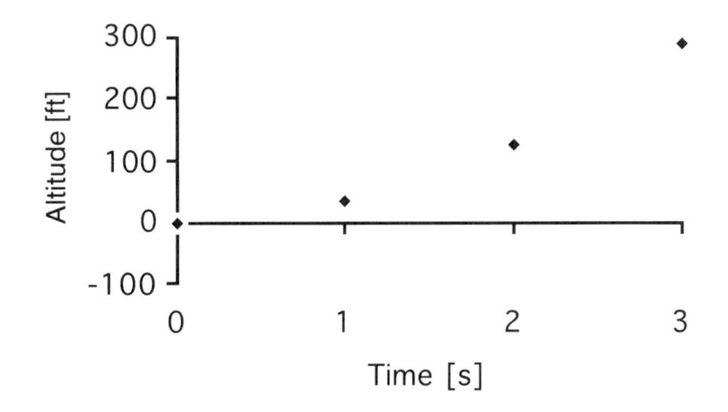

Fig. E.3. Idealized Altitude vs. Time for a miniature rocket

This graph looks a bit like one of the parabolas we all learned about in algebra classes. The general equation for a parabola is given by the expression

$$y = ax^2 + bx + c$$

where a and b are constants. If we were to attempt to find an equation that seems to describe the altitude of our rocket as a function of time, we might guess that the expression would be something like

$$\text{Altitude} = a_2 \times \text{Time}^2 + a_1 \times \text{Time} + a_0$$

since time can be assigned as the independent variable plotted along the x-axis and the altitude can be assigned the role of the dependent variable plotted along the y-axis. To do mathematical modeling you would need to plot the theoretical data for different values of a_1, a_2, and a_0 until a theoretical curve and the experimental curve match pretty well. The process can be simplified a bit because someone with experience would recognize immediately that a parabola of the type that fits the rocket data would have $a_1 = 0$ and $a_0 = 0$. Thus, only a "best" value for a_2 is needed.

Suppose you graphed some other data that doesn't look like either a parabola or a straight line. and you don't know the exact mathematical relationship that can be used to fit your data. For example, a graph of the temperature of a cup of coffee that is cooling over time does not look like either a parabola or a straight line. In that case you need to guess the functional form of your curve using a knowledge of physics theory or a famialiarity with the shapes of common functions. You might try powers of x; or e^x, or perhaps $\ln(x)$ in some cases.

MATHEMATICAL MODELING WITH AN EXCEL SPREADSHEET

Unfortunately, it is very tedious to guess an equation because after guessing the form of the equation, you also need to find good values for the constants in the equation and then plot graphs for each guess. Fortunately, an Excel worksheet with its dynamic graphing capability can be used to make the modeling process much easier.

An Excel Modeling Tutorial (MODTUT.XLS) has been developed to help you learn about the process of spreadsheet modeling. Each time you perform simple one-dimensional mathematical modeling, you can open a Worksheet file and set it up so that one column of data represents the experimental data points and another column of data represents the proposed theoretical data. These columns can be linked to a graph. The first time you attempt to do mathematical modeling you should open (MODTUT.XLS) the *Modeling Tutorial*. In this tutorial various notes and messages have been included to help you get started with the modeling process.

A portion of the *Modeling Tutorial* worksheet is shown in Fig. E.4. Open up the *Modeling Tutorial* spreadsheet and enter a mathematical relationship in the first cell in the y-theory column that fits the sample experimental data given.

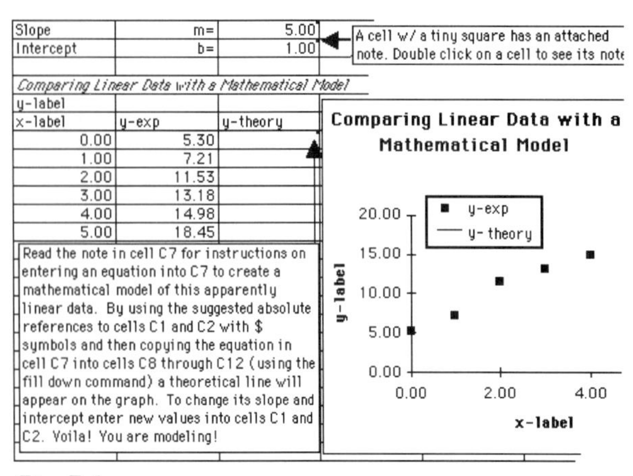

Fig. E.4.

After modeling the data in the tutorial spreadsheet, you can open a new worksheet and try to model the data shown in Figure E.1 describing the horizontal distance a pitched baseball moves over time. Do you agree that the mathematical equation is a linear one? Is 19.5 m/s a good estimate for the slope?

BECOMING FAMILIAR WITH CURVE SHAPES FOR DIFFERENT FUNCTIONS

If data seems to lie along a certain type of curve or if a theory suggests that data ought to lie along a certain curve, then you can make an intelligent guess about the equation needed for a mathematical model. To make such a guess, it is helpful to know what the shapes of curves associated with various mathematical expressions are like. For example, what does $y = A\sin(Bx)$ look like for various values of A and B? Creating graphs of various types of functional relationships is a good way to begin to recognize the general shapes of curves. A special Excel toolset called the WPtools for Excel has been created to allow you to produce graphs of various functions easily.

Using the Graphing Tool to Display Common Functions Graphically:

You can use the Excel graphing tool to generate a graphical library of mathematical relations that frequently occur in physics. Some functions that can be plotted include:

1. $y = m \times x + b$
2. $y = x^2$
3. $y = x^P$ where $P = 1, 2, 3 \ldots 98$
4. $y = e^x$
5. $y = 1/x^P$ where $P = 1, 2, 3 \ldots 98$
6. $y = 1/e^x$
7. $y = x$
8. $y = \mathrm{SQRT}(x)$
9. $y = \ln(x)$
10. $y = A\sin(B \times x)$ for $A = 1, 2, 3$ and $B = 1, 2, 3$, etc.

To use the graphing tool, open an Excel *Worksheet* and enter values in an x column and values in a y column generated by the mathematical equation of interest. (See Appendix A for details on using the WPtools for Excel for Graphing.)

Let us carry out a systematic investigation of several sequences of functions.

a. Turn on a your computer and open an Excel Worksheet file.

b. Title the first colomn x, and enter about 20 equally spaced values of x of interest to you into the x-raw column. For example, if you are interested in functions of x where x varies between 0 and 10 you could make a column of numbers such as 0.0, .5, 1.0, 1.5 . . . 10.0.

c. Title the second colomn y. Select the first cell in the y column. Then choose the formula you wish to use to transform your data. You may choose to take the natural log of your x-data, or see what happens when you raise it to a power. Once you've entered the formula or function of your choice, you can Fill Down into the rest of the y-raw column.

d. Finally, plot your raw data by selecting your x and y data columns and clicking on the graphing tool that is on the left-hand side of the custom toolbar at the bottom of the screen.

You should carry out a systematic investigation of several sequences of functions using the graphing tool. To record results, you should sketch the shape of each graph in your Lab Notebook or Activity Guide. Please label the axes for each graph carefully and note how the curves change as you proceed through a sequence of functions.

1. *Rapidly growing curves:* Try the functions numbered 2 to 4, and note how they become more "abrupt" in their departure from the x-axis.

2. *Rapidly dying curves:* Try the functions numbered 5 and 6.

3. *Slowly growing curves:* Finally, try function numbered from 7 to 9. Note that the logarithm approaches negative infinity near $x = 0$.

All these relationships occur in some branch of physics. We hope you will find this exercise useful in later mathematical modeling activities. Indeed, now that you know how to use the WPtools for Graphing, you can call up your graphical library of functions at will. If you can guess the functional form of a physical relationship by viewing its graph, you can often develop a mathematical model to describe it.

APPENDIX F UNCERTAINTY PROPAGATION–UNCERTAINTIES AFTER CALCULATIONS

Maids to bed and cover coal;
Let the mouse out of her hole;
Crickets in the chimney sing;
Whilst the little bell doth ring:
If fast asleep, who can tell
When the clapper hits the bell?

UNCERTAINTIES COMPOUND

Imagine what would happen if you were trying to determine the area of a rug using an old, warped stick marked in centimeters from 0 to 300. You find the length of one side of the rug to be 200 cm, but since the measuring stick could have shrunk or expanded, you attach a 5-mm or 0.5-cm uncertainty to your measurement and write down 200 ± 0.5 cm for the length of one side. You make a mental note that this is a $(0.5/200) \times 100\% = 0.25\%$ percentage uncertainty. Now you go to measure the other side, and, lo and behold, you find that the rug apears to be square—you get 200 cm again, but you attach the same uncertainty to it: 200 ± 0.5 cm.

Let's say you want to sell this rug to a wise old Armenian rug merchant, who asks you exactly how certain you are of its area. We will draw a diagram of the situation below.

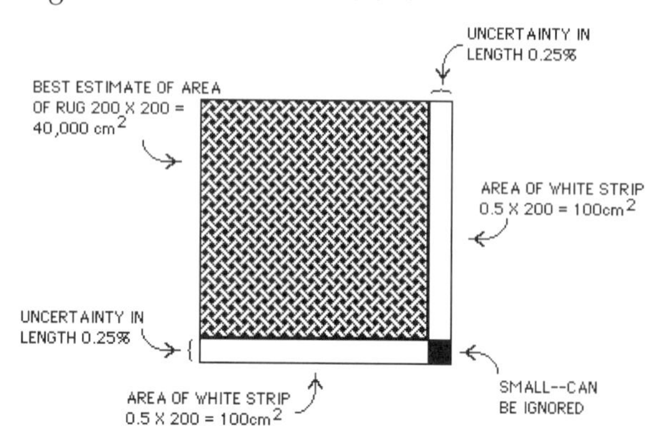

Fig. F.1.

The picture shows the rug with the best estimate of its dimensions, 200 × 200 cm. It also shows as white strips (greatly exaggerated in size!) the "extra area" that would result if each side was increased in length by the full amount of the uncertainty in the measurement of the length of a side. These two white strips contribute an extra area of 200 cm^2. So the rug's area is uncertain by that amount. The area should be quoted as 40,000 ± 200 cm^2. We are ignoring the small black overlap region of area (0.25 cm^2) since it is the square of an already small uncertainty in length.

Now comes the important point: the fractional uncertainty of the area of the rug is $(200/40{,}000) \times 100\% = 0.5\%$. We have gone from a 0.25% uncertainty in the length of a side of the rug to a 0.5% uncertainty in the rug's area! Although the wise old Armenian rug merchant might not complain, things are worse by a factor of two!

The point of this parable is this: *when numbers that are themselves uncertain are used in a calculation to obtain a final result (squaring the length of one side of a square to get an area, for example) the final result is very often more uncertain, in percentage terms, than the original numbers were!*

More specifically, if

$$y = x^2$$

then it can be shown that

{percentage uncertainty in y} =
 $2 \times$ {percentage uncertainty in x}

We will outline the general theory behind all this, but you should keep in mind that percentage uncertainties tend to increase as new quantitatives are calculated. Frequently we start with measurements that we think are fair to good, but we can be unpleasantly surprised by the uncertainty in the final calculated result!

THE PROPAGATION OF UNCERTAINTIES

More often than not, the final result of an experiment involves the measurement of several numbers that are then combined in some formula. How do the uncertainties in each of these numbers combine to produce the uncertainty in the final result? The answer for the general case is a bit complex, but if we can assume that the uncertainties in each of the "input" numbers in the formula are unrelated (as is often true), we can use the simple rules that will be discussed below. *Unrelated uncertainties* means that the size of the uncertainty in one quantity does not directly affect the size of the uncertainty in any of the other input quantities.

Let us assume that our final result is $R = x + y + z$, and that we can estimate the uncertainties in x, y, and z. Then we have the following relationship between the uncertainty in the calculated result and those in x, y, and z:

σ_x = Uncertainty in x
σ_y = Uncertainty in y
σ_z = Uncertainty in z
σ_R = Uncertainty in R

$$\sigma_R = \sqrt{\left(\sigma_x^2 + \sigma_y^2 + \sigma_z^2\right)}$$

If, instead, $R = x + y - z$, the above rule still holds. In general, we say that for results involving sums and differences, uncertainties add "in quadrature," meaning as a sum of squares.

If our calculations involve products and quotients, the method for determining the resulting uncertainty is more complex. In such a case it turns out that the fractional uncertainties, defined as the uncertainty divided by the best estimate of the value, add in quadrature. For example if the result is the product of x and y divided by z, so that $R = x\,y/z$ then:

$$\frac{\sigma_R}{R} = \sqrt{\left(\frac{\sigma_x}{x}\right)^2 + \left(\frac{\sigma_y}{y}\right)^2 + \left(\frac{\sigma_z}{z}\right)^2}$$

Another common situation is when R is a function of some one variable, $R = f(x)$. Then:

$$\sigma_R = \left|\frac{df}{dx}\right| \sigma_x$$

Here the vertical bars indicate "absolute value"—so we are ignoring the sign of the derivative. The derivative is evaluated at the best estimate of x. A simple example of this is when the result is the product of a constant A of x so $R = A_x$. Then the above formula tells us that the uncertainty in R is A times the uncertainty in x:

$$\sigma_R = \left|A\right| \sigma_x$$

A more complex example would be the case where $R = \ln(x)$, where ln indicates the natural logarithm. Since the derivative of the natural logarithm is $1/x$, our rule then becomes

$$\sigma_R = \left|\frac{1}{x}\right| \sigma_x$$

We have presented these rules without derivation, but we hope that you find them at least plausible. An excellent discussion of these and many other related topics that we cannot hope to cover in this Appendix is given in the book *An Introduction to Error Analysis—The Study of Uncertainties in Physical Measurements* by John R. Taylor (University Science Books, Mill Valley, CA, 1982). The interested student is strongly encouraged to refer to this little book.

APPENDIX G
THE METHOD OF LEAST SQUARES ANALYSIS

OVERVIEW

In both Appendix E and in the first few Workshop Physics Activity Guide Units you have been taught to develop analytic mathematical models of data that you expect to lie along simple curves. In most cases you have been asked to match your "best fit" function to data by eye. Your "best" curve or line is the one that seems to be the "closest," on average, to all of the points.

Mathematicians have developed more exact techniques for *fitting* functions to data. The most common of these techniques is called the method of Least Squares. In this method the criterion for "best" fit is stated as follows: *the sum of the squares of the vertical distances between each data point and the best fit curve or line is a minimum.* In this appendix we will: (1) outline the derivation of the equations used to find values for *slope* and *intercept* from a set of linear data; (2) discuss how the linear fit tool from WPtools can be used to fit linear data; (3) explain how the same approach can be applied to fitting data points lying on a curve with a polynomial function; and (4) learn how to use the WPtools polynomial fit tool to fit data with a polynomial function.

LEAST SQUARES FOR LINEAR DATA

The least squares method of fitting experimental data to a theoretical curve is one of the most frequently used data analysis procedures in experimental science and engineering. In the most elementary applications of the method, one of the coordinates of each point is assumed to be known exactly, i.e., it is entirely free from uncertainties. You should be aware that this is sometimes a poor assumption. The 'uncertainty free' coordinate is typically chosen as the x coordinate so that the experimental uncertainty in each data point (x_i, y_i) resides in only y_i. Figure G.1 shows a set of experimental data points through which a least squares best straight line has been drawn.

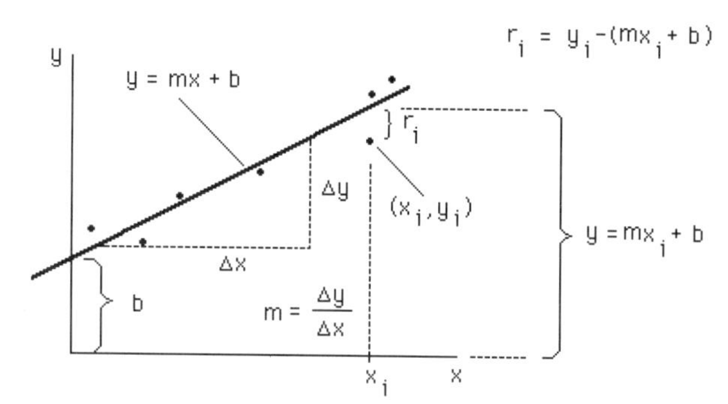

Fig. G.1.

The difference between a measured y_i and the height of the line at $x = x_i$ is called the residual. Clearly, the height (ordinate) of the line at $x = x_i$ must be $mx_i + b$, which is defined as the residual, r_i, for the ith point from the least squares line. The residual equation that is used to compare each is shown in equation G.1.

$$r_i = Y_i - (mX_i + b) \qquad \text{(G.1)}$$

DERIVATION OF THE EQUATIONS FOR *m* AND *b*

We want to find the slope, m, on the intercept; b, m and b for which the sum of the squares, S, of the residuals for all n data points in the data set is a minimum. That is, we select the m and b for which

$$S \equiv \sum_{i=1}^{n} r_i^2 = \text{a minimum} \qquad \text{(G.2)}$$

We can begin to derive the equation to find the best fit values of m and b by substituting equation G.1 into equation G.2 and squaring equation G.1.

$$S \equiv \sum_{i=1}^{n} r_i^2 = \sum_{i=1}^{n} [Y_i - (mX_i + b)]^2 \qquad \text{(G.3)}$$

$$= \sum_{i=1}^{n} (Y_i^2 + m^2X_i^2 + b^2 + 2mbX_i - 2mX_iY_i - 2bY_i)$$

$$= \sum_{i=1}^{n} Y_i^2 + m^2\sum_{i=1}^{n} X_i^2 + nb^2 + 2mb\sum_{i=1}^{n} X_i - 2m\sum_{i=1}^{n} X_iY_i - 2b\sum_{i=1}^{n} Y_i$$

To find the equations that give a best estimate of m and b, the value of the sum of the squares of the residuals must be minimized. To do this the following conditions must be met:

1. The partial derivative of G.3 with respect to m is zero.

$$\frac{\partial S}{\partial m} = 0$$

2. The partial derivative of G.3 with respect to b is zero.

$$\frac{\partial S}{\partial b} = 0$$

Note: A partial derivative of a function of several variables with respect to one of the variables is calculated by treating all the other variables as if they were constants and differentiating in the usual way.

Recall that the values of X_i and Y_i represent data and therefore are constants for this problem. If our only variables are m and b, the first condition yields equation G.4.

$$\frac{\partial S}{\partial m} = 2\sum_{i=1}^{n} X_i^2 + 2b\sum_{i=1}^{n} X_i - 2\sum_{i=1}^{n} X_iY_i = 0 \qquad \text{(G.4)}$$

The second condition leads to equation G.5.

$$\frac{\partial S}{\partial b} = 2nb + 2m\sum_{i=1}^{n} X_i + 2\sum_{i=1}^{n} Y_i = 0 \qquad \text{(G.5)}$$

These two equations may be solved for m and b, yielding:

$$m = \frac{n\sum_{i=1}^{n} X_iY_i - \sum_{i=1}^{n} X_i \sum_{i=1}^{n} Y_i}{n\sum_{i=1}^{n} X_i^2 - \left(\sum_{i=1}^{n} X_i\right)^2} \qquad \text{(G.6)}$$

$$b = \frac{\sum_{i=1}^{n} X_i^2 \sum_{i=1}^{n} Y_i^2 - \sum_{i=1}^{n} X_i \sum_{i=1}^{n} X_iY_i}{n\sum_{i=1}^{n} X_i^2 - \left(\sum_{i=1}^{n} X_i\right)^2} \qquad \text{(G.7)}$$

A careful analysis including taking the statistical variances of these expressions yields confidence intervals for m and b. You should not be surprised to learn that these confidence intervals are related to the sum of the residuals squared.

LINEAR LEAST SQUARES ANALYSIS WITH THE WPTOOLS

Suppose you want to find the "best" values, in the least squares sense, of slope, m, and intercept, b, that correspond to a graph of data points that lie on or close to a straight line. An example is shown in figure G.2. The data has been entered into an Excel worksheet with the Plot Title, the Independent and Dependent variable labels, and their values entered in the locations required by the WPtools. (See Appendix A or the WPtools Manual for details on the custom WPtools.)

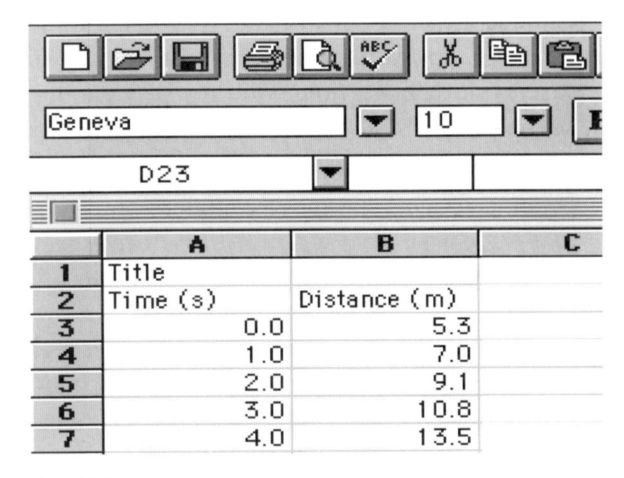

	A	B	C
1	Title		
2	Time (s)	Distance (m)	
3	0.0	5.3	
4	1.0	7.0	
5	2.0	9.1	
6	3.0	10.8	
7	4.0	13.5	

Fig. G.2.

Suppose, after closely looking at your data (possibly using a scatter plot), you suspect that:

$$\text{Distance} = \text{Slope} \times \text{Time} + \text{Intercept}$$

so that the equation relating Distance and Time

has the general form $y=mx+b$ (or $y=a_1x+a_0$ in the notation used by the WPtools) where $m=a_1$ represents the slope of the line and $b=a_0$ its intercept. You can obtain a linear least squares fit by selecting the data with their column labels and then using the custom WPtools Linear Fit tool. The Linear Fit tool is available on the *WPtools* menu or on the *WP Standard* toolbar. The toolbar icon looks like a scatter plot with a line through it as shown below. (See Appendix A for details on the custom WPtools.)

a. Create x-data, and y-data columns and copy your data into them with the independent variable in the x-data column and the dependent variable in the y-data column. Your entered data should have a format like the data shown above.

b. Select the dependent and independent variables to be fit. Be sure to select the independent variable column first.

c. Locate the custom *WP Standard* tool bar and Click on the *Linear Fit Tool.*

OR

Choose *Linear Fit* from the *WPtools* menu. Wait for a minute or so while the computer carries out calculations similar to those described in the *Least Squares for Linear Data* section of this Appendix.

d. When the calculations are complete, the WPtools Linear Fit Tool will add a new worksheet ply to the active workbook containing the results of the fit. This new sheet will have the same name as the sheet containing the data that was entered in step a with the text "Fit #" appended, where # represents a unique number.

If a default installation of the WPtools is being used, a plot as shown in Figure G.3, which contains all of the fit information will be added to your original worksheet. (See the manual for the WPtools to learn how to customize the output of the fit tools.)

e. Since we are postulating that Distance = Slope × Time + Intercept, the constant fit coefficient, $a_0=5.10$, represents the best fit for the intercept. The best fit for the slope is represented by fit coefficient $a_1=2.02$.

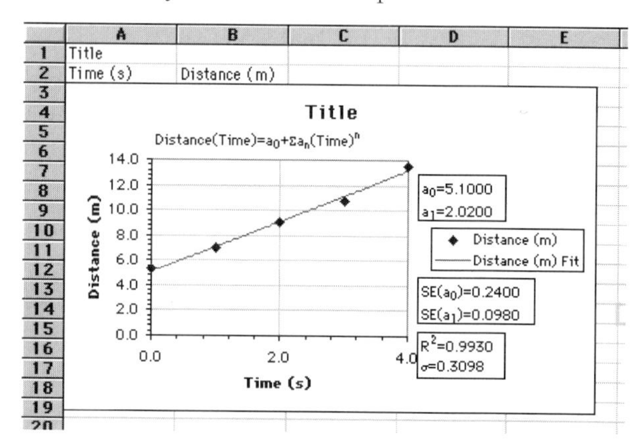

Fig. G.3.

All the fit information is presented in the text boxes on the graph. These include:

1. Directly above the graph and below the title is the equation for the fit line in terms of the independent and dependent variables.
2. Above the legend to the right of the graph is a box that contains the values of all of the fit coefficients. In this case the intercept, a_0, and the slope, a_1, of the fit line are shown.
3. Just below the legend is a box that contains the standard errors (S.E.) of each of the fit coefficients. The S.E. for the slope is given by SE(a_1), and likewise the S.E. for the intercept is given by SE(a_0).

Assuming a sufficiently large data set and a normally distributed experimental uncertainty, the standard errors represent approximately a 68% confidence interval. This indicates that the true value of the slope (and intercept) of the process from which the data set was taken has about a 68% chance of lying within one S.E. of the slope (and intercept) found by the least squares analysis. Note that with data sets that are less than infinitely large, the actual 68% confidence interval will be larger than indicated by the standard error. The actual size of the confidence interval involves the use of the *t distribution* from statistics.

4. Below the standard error box is the fit statistics box which contains the square of the correlation coefficient, R^2, and the standard error about the regression, σ.

The value of the square of the correlation coefficient, R^2, is an important indicator of the validity of your fit equation. The R^2 value reflects the proportion of the variations in the measured data values about their means that are attributable to the experimental uncertainty, which is assumed to be normally distributed. The remaining portion of the variation in the measured data is due to the inability of the fit equation to accurately describe the data. Therefore, as the value of R^2 approaches unity all of the variation in the data can be attributed to the experimental uncertainty and the fit equation is considered to be correct. If, however, R^2 is close to zero, the variations are due primarily to the fact that the data is not accurately described by the fit equation and the use of that particular fit equation must be reconsidered. Very simply, an R^2 value near 1 indicates an excellent fit and values of R^2 near zero indicate a poor fit.

The standard error about the regression, σ, represents approximately a 68% confidence interval for the data at the mean value of the independent variable. In other words, the true value of the data at the mean of the independent variable has a 68% chance of being within one σ of the value predicted by the fit. Note that there is a tacit assumption here that the experimental uncertainty is normally distributed and the fit equation used is correct (see discussion of R^2 above). Also note that, for data sets less than infinitely large, as you move away from the mean of the independent variable the size of the 68% confidence interval will increase. The rate of increase is affected by the distance from the mean of the independent variable and the number of data points used in the regression.

POLYNOMIAL LEAST SQUARES ANALYSIS

An Example of Data That Can Be Fit by a Polynomial

If the data points you want to fit lie on a curve rather than a line, the equation that fits the data might be polynomial of the form

$$y = a_n x^n + a_{n-1} x^{n-1} + \ldots + a_1 x + a_0 = a_0 + \sum_{i=1}^{n} a_i x^i$$

where the ai's are constants. The highest power in the polynomial determines its order. The equation above is for a polynomial of order n. **Note:** In the Least Squares for Linear Data section, the variable n represented the number of data points. Here, n is used to indicate the order of the polynomial fit equation.

Several questions that arise when least squares analysis is used for polynomial functions are:

1. For a given curve that is actually a polynomial, what is the best value for each a_i?
2. How good is the notion that a polynomial is the right kind of relationship?

Let's consider a different set of rocket data similar to that presented in Appendix E in which the altitude of a rocket is measured as a function of time. A data summary is shown in Figure G.4.

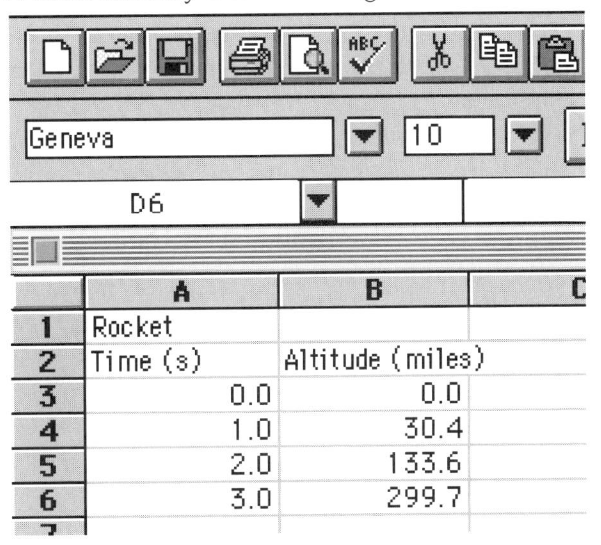

	A	B	C
1	Rocket		
2	Time (s)	Altitude (miles)	
3	0.0	0.0	
4	1.0	30.4	
5	2.0	133.6	
6	3.0	299.7	

Fig. G.4.

A graph of these data points indicates that the relationship between time and altitude lies on what looks like a polynomial of order 2. That's a fancy way of saying the "curve" looks like a parabola.

It is possible to use least squares analysis with polynomial and other types of functions. For example, from the graph in Figure G.5, it appears that the equation relating altitude to time might be a polynomial of order 2 with the form:

$$Altitude(\text{mi}) = a_2 time(\text{s})^2 + a_1 time^1 + a_0$$
$$= a_0 + \sum_{i=1}^{2} a_i time(\text{s})^n$$

where a_2 is the coefficient of the time squared term, a_1 is the coefficient of the time term, and a_0 is the constant term (or the coefficient of the time^0 term).

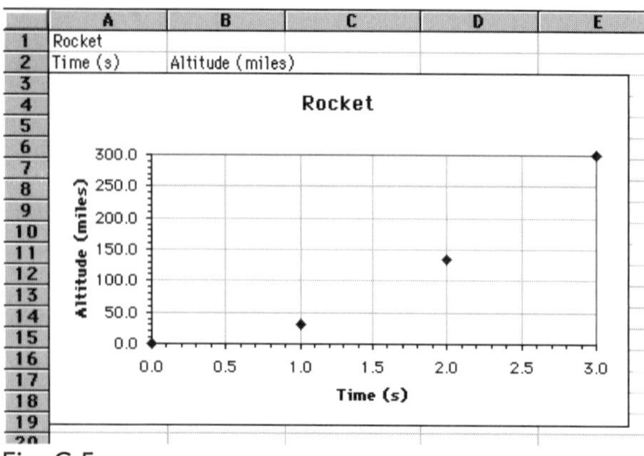

Fig. G.5.

Mathematicians have derived equations that can be used to find the "best" values of various coefficients representing polynomials. Since real data points rarely lie exactly on a curve, it is important to be able to estimate the uncertainty associated with a given fit. A measure of uncertainty used in least squares analysis is the standard error (S.E.) of each of the coefficients (see discussion in the section on *Linear Least Squares Analysis with the WP Tools*).

Using the WPtools Polynomial Fit Tool

If you want to find the "best" values of the coefficients of a polynomial that correspond to a graph of data points that lie more or less on a curve, you can obtain a polynomial least squares fit using the Polynomial Fit Tool from the WP-tools. This tool looks like a scatter plot with a curve through it as shown below.

a. Open an Excel 5.0 workbook.

b. Once the workbook is open, enter the independent and dependent variable data into columns and give the data columns appropriate titles.

c. Highlight the data pairs that you want to fit. Include the data titles in the highlighted area. **Note**: If the columns are not together with the dependent variable on the left, be sure to highlight the independent variable column first.

d. Locate the custom *WP Standard* tool bar and click on the *Polynomial Fit Tool.*

OR

Choose *Polynomial Fit . . .* from the *WPtools* menu.

After few seconds a dialog box should appear asking you what order of polynomial you want to fit. Fill in your best guess about the integer that describes the order of the polynomial (i.e., 1 or 2 or 3, etc.) and click the OK button. If you are an advanced user you may select the advanced options for more flexibility.

Wait patiently for a minute or so while the computer carries out calculations which are extensions of those described above in the *Least Squares for Linear Data* section.

e. When the calculations are complete, the WP-tools Linear Fit Tool will add a new worksheet ply to the active workbook containing the results of the fit. This new sheet will have the same name as the sheet containing the data entered in step 1 with Fit# appended, where # is a unique number.

If a default installation of the WPtools is being used, a plot that contains all of the fit information will be added to your original worksheet. An example is shown in Figure G.6.

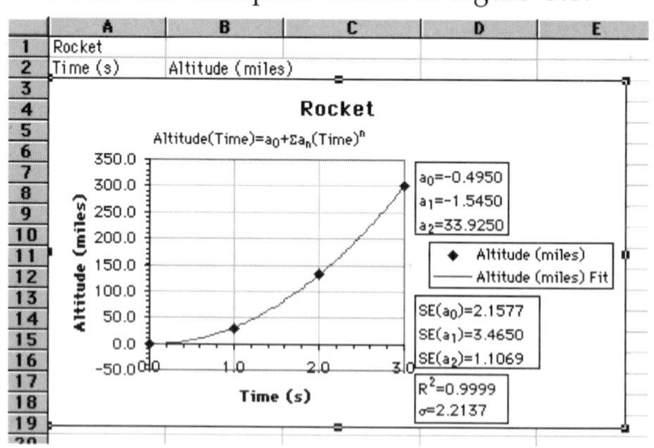

Fig. G.6.

(See the manual for the WPtools to learn how to customize the output of the fit tools.) The significance of each of the fields on the plot is discussed above under Linear Least Squares with the WPtools.

APPENDIX H
VIDEOPOINT SOFTWARE

AN INTRODUCTION TO VIDEOPOINT

VideoPoint in a Nutshell

The VideoPoint software allows you to collect coordinate data by clicking with a mouse on locations of interest on video images. This allows you to study two-dimensional motions by locating, displaying, and analyzing coordinate data obtained from sequences of digitized video frames. You can also study individual electronic images saved as QuickTime movies to determine geometric relationships or count objects of interest. The VideoPoint software can be operated from either menus or a toolbar. It has drag and drop features and a balloon-like help system. VideoPoint runs on both the Windows and Macintosh platforms.

For Those Who don't Read Manuals. . .

If you are not a manual reader, the fastest way to get started is to read the entries on "video points" and "coordinate systems" in part 1.3 of this Chapter and then skip to Chapter 2 on Getting Started.

An Introduction to VideoPoint Features

The VideoPoint software allows you to define characteristics of a series of points you would like to locate on each frame. These point series characteristics include its name and the size and shape of its markers. Other point series characteristics that can be determined are the mass of the object being located and the coordinate system each point series is associated with. You will also be able to specify the length of objects or distances between features in frames for scaling purposes. In addition to obtaining data via the selection of features or objects of interest on frames, you will be able to define calculated data points series such as the locations of the centers of mass of an object. Data that are obtained can be graphed as they are located or calculated. Data can be saved in an electronic file or copied for use with other types of analysis software such as spreadsheets and graphing programs.

Conventions Used in this Manual

Menu Choices: File->Open Movie... is equivalent to "choose 'Open Movie...' from the 'File'-menu".

! Important Note. Common errors and information on features that aren't obvious.

? Common Question and Answer. These answer some of the questions that arise concerning certain features and methods.

Macintosh instructions: The abbreviation MAC is used to denote specific instructions for the Macintosh computer.

Windows Instructions: The abbreviation WIN is used to denote specific instructions for a Windows PC.

Screen Shots: Almost all of the screen shots depicted in this manual are from the Macintosh version. The screens shown are functionally equivalent to the Windows version.

BASIC INFORMATION

"Video Points"

A video point is defined as a location of a feature or object of interest on a single movie frame. The software initially stores the (x, y, and t) values of a video point; where x is the distance from the left side of the movie window (in pixels); y is the distance from the bottom of the movie window (in pixels); and t is the elapsed time in seconds since the first frame in the movie was recorded.

By themselves, video points are not very interesting. However, the VideoPoint software allows you to make calculations based on these video points.

What is the difference between a video point and VideoPoint? The term VideoPoint refers to the name of the software while the term video point refers to a point you have located on the frame of a QuickTime movie.

Video points are designated by you. For example, if you are looking at a movie of a ball toss, you might be interested in measuring the positions of the ball for the entire toss. In order to do this, you would set up the VideoPoint software for one point per frame, and then click on the location of the ball in each frame in the movie. VideoPoint then stores the information for the series of video points corresponding to the selected locations.

Since the data set, consisting of a series of video point coordinates, is stored in screen units (pixels) and is relative to the arbitrary origin of the bottom left of the movie, it isn't terribly useful for analysis. Thus, you have the ability to define various coordinate systems. You can then associate the video points with a coordinate system and determine the position coordinates in the system they are associated with.

Coordinate Systems

The VideoPoint coordinate system is two-dimensional and consists of an origin, an orientation, and an optional scale factor. In addition, you can designate a coordinate system as either Cartesian or polar. By default, VideoPoint opens a movie with two coordinate systems present. The first coordinate system is known as the default system and is initially named the "Origin 1" system. It is a Cartesian system with horizontal and vertical axes and a preselected origin (i.e., Origin 1) near the lower left of the movie window. Initially the units of the coordinates in this system are in pixels. You can easily change the name of this system or scale it so that video points you locate have coordinates in meters or centimeters. You can also move the default system origin and rotate the coordinate axes if you choose. The second coordinate system is VideoPoint's native system, the "Video Origin" system. This is a Cartesian coordinate system with horizontal and vertical axes and no scaling. The coordinates of the video points located in this system are always in pixels and the "Video Origin" is always at the bottom left of the movie. You cannot change the "Video Origin" system in any way.

Each video point series that you define has to be associated with a coordinate system. Video points that are associated with "Origin 1" coordinate system have (x,y,t) data saved as coordinates in the "Origin 1" coordinate system.

Scale Factors

Data stored in pixels is only useful for computers. In order to collect data in "real" units (i.e, meters), each coordinate system must be scaled. In a sample movie, a meter stick might appear to be about 200 pixels tall. During the scaling process, you need to click on both ends of this meter stick and tell the VideoPoint software that the distance between these two video points (which VideoPoint sees as 200 pixels) is actually 1.00 meters. VideoPoint would then assign a scale factor of 200 pixels/m. You can then associate the scale factor with any of the coordinate systems you have defined. With the combination of the origin location and the scale factor, video point data can be reported in "real" units relative to any coordinate system.

Calculations Based on Video Points

You can specify the standard calculation based on two coordinates or two or more video points associated with a given coordinate system. Each of these calculated items is described in depth in the "How Do I?" chapter. These calculations include:

Distance: The distance between any two video points on a frame.

Scale: Ratio of a known length (in meters or centimeters) to the distance in pixels between two video points.

Center of Mass: Calculated center of mass of a collection of video points based on masses associates with a series of video points. Each series of video points can be assigned a different mass.

Angle: Angle made by lines connecting three video points.

Designated Point: Point at a location specified by relative distances between any two video points

Movies

Movies are sequences of still images that have been digitized and saved in the QuickTime™ format. Each image is called a frame. Each frame has a time associated with it that represents the elapsed time since the first frame of the movie was recorded.

Where do Movies come from?

A collection of over 260 digital movies can be found on the VideoPoint CD. These sequences of video frames were recorded using a standard video camera and then later digitized by a computer and transformed into a popular computer format known as QuickTime. With appropriate computer hardware installed in your Macintosh or PC computer, you can also create your own QuickTime movies that can be analyzed with VideoPoint.

VideoPoint Files

Information derived by VideoPoint can be saved in several ways. The entire data set along with the current window arrangement (including movies, data tables, and graphs) can be saved as a VideoPoint file with the extension .VPT.

Windows computers: VideoPoint files have the extension .VPT

Macintosh computers: Though the sample files all comply to the DOS file naming convention, you can use any name under 32 characters in length.

VideoPoint files contain coordinate data as well as the name and location of the movie. It does NOT save the movie itself in the file, nor does it ever edit the movie. Thus, if you want to open a VideoPoint file later, the movie associated with this file must also be present. The only exception is when you save a template for later analysis of similar movies.

The data can be copied from the data table into the clipboard and spreadsheet or any other program that accepts a tab-delimited format. The entire data set can also be exported as a tab-delimited text file.

A TYPICAL ANALYSIS

The easiest type of movie to analyze is a movie with a single feature of interest that was taken with a camera that does not move or zoom during filming. Let's start by analyzing a ball launched from a fixed table.

Opening VideoPoint

Macintosh computers: Double-click on the icon in the finder.

Windows computers: Double-click on the Program Item in the VideoPoint group.

SETUP SCREENS

Once VideoPoint is opened, the following setup screen will appear. To start, click on "Open Movie . . ." to open the movie you want to analyze. To work with this example you should choose the movie entitled "PRJCTILE.MOV" located in the samples directory.

Fig. H.1. The VideoPoint title screen.

A screen showing the first frame of the movie should appear. You will be invited to enter the number of features or objects that you want to locate on each frame. Since only one ball is launched during this movie, type 1 into the box and hit "OK."

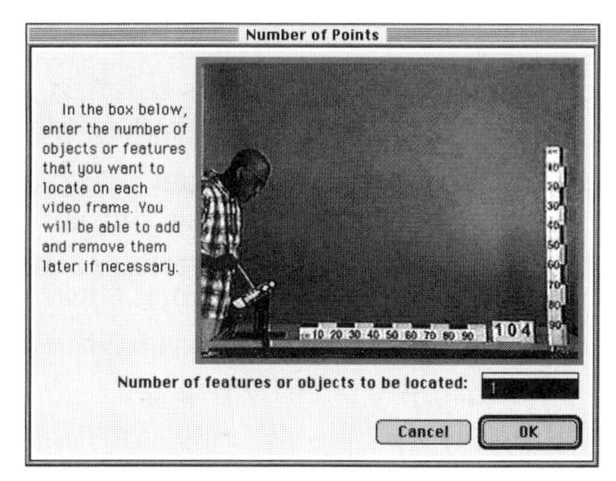

Fig. H.2. The second screen, which allows you to indicate the number of features or objects of interest to be located on each frame of the movie.

The movie should appear on the right along with a coordinate system window on its left as shown in Fig. H.3. Initially the coordinate system window shows two Cartesian systems; the "Origin 1" coordinate system and the "Video Origin" coordinate system. A single video point series (the one specified in the previous dialog box) called "Point s1" has been placed in the "Origin 1" coordinate system; "Point s1" will report all its data relative to the "Origin 1" coordinate system.

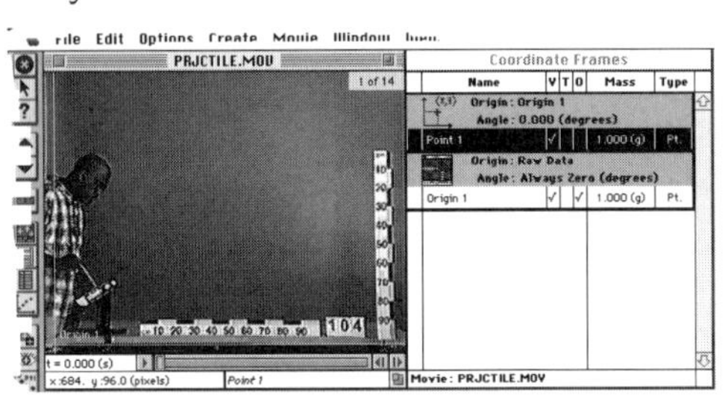

Fig. H.3. The default working screen showing the movie, coordinate frame, and data table windows.

PLAYING THE MOVIE

Play the movie to see what happens by clicking on the play button on the movie controller, located at the bottom of the screen.

Rewind the movie by either dragging the slider on the movie controller back to the beginning of the controller or by choosing "Movie→Rewind" (Ctrl-R).

MAC users: Use Command-R instead of Ctrl-R.

TAKING DATA

Move the cursor over the movie window area. The cursor should look like the reel, and the bottom right of the movie window should have the italicized text *"Point s1."* This is the first video point to be selected by you in the frame that currently appears in the movie window.

Cursors in the Movie Window:

If the cursor looks like this, clicking on the movie window will locate the video point that is currently selected.

If the cursor looks like this, you can select a previously located video point and use a drag and drop technique to move it to another location.

Move the cursor so that the ball being launched out of the projectile launcher is centered in the cursor. Click once to locate this video point.

The movie will automatically advance to the next frame. Continue clicking on the location of the ball in each frame until the last frame of the movie.

You have now collected data for this movie. However, the data is still in pixels since you have not yet scaled the movie.

SCALING THE MOVIE

This process tells VideoPoint how many screen units (pixels) in the movie window are in a meter, a centimeter, or a millimeter in the actual scene. Conveniently, a 1.00 meter long meter stick was placed in "PRJCTILE.MOV"; this will be used to scale the movie.

STARTING THE SCALING PROCESS

Click on the scale icon in the toolbar. The following dialog will appear.

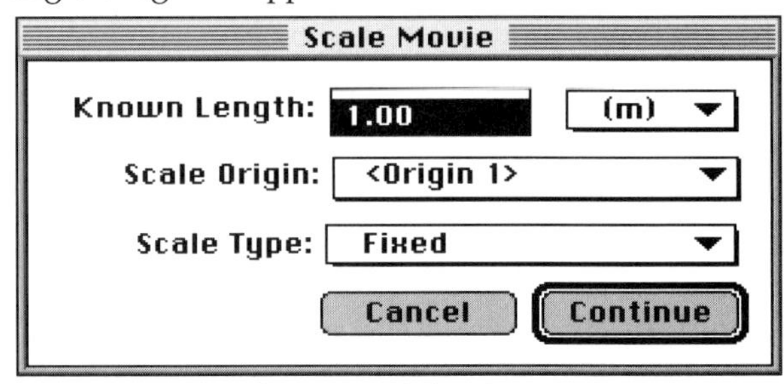

Fig. H.4. The Scale Movie dialog box.

The length of the scale object (in this case, the meter stick) is known to be 1.00 meter. Enter this value into the "Known Length" box. Since we want to scale the coordinate system relative to "Origin 1," select "Origin 1" in the "Scale Origin:" pop-up menu. Since the camera did not zoom at the instant when the movie was taken, choose a "Fixed" scale type.

Once these values have been set, you are ready to begin the scaling process.

a. Click on "Continue."

b. Click once on one end of the meter stick.

c. Click once on the other end of the meter stick.

Note that three new rows appear on the Coordinate System's Window. Two rows, Scale 1A and

Scale 1B, specify the ends of the object that you clicked on. The third line, called Scale 1, stores the ratio of the length of the object relative to the distance between Scale 1A and Scale 1B.

Now you have scaled this coordinate system by telling VideoPoint that 1.00 meter is equivalent to the distance (in pixels) between the two points that you just clicked on.

> What are the Scale lA and Scale lB video points? These two video points are used by the program to determine the number of pixels between the ends of an object or the distance between two features used for scaling on a video frame. If the actual distance (in meters, centimeters, or millimeters) between the Scale lA and Scale lB video points is known, then a scale factor can be determined for the frame. This scale factor is calculated as the ratio between the number of pixels between Scale lA and Scale lB and the actual distance between these points specified by you. Moving either of the two scaling video points closer together will decrease the scale factor and moving them farther apart will increase the scale factor.

GRAPHING DATA

To graph the data that you have collected, click on the graph icon in the toolbar or choose *View→New Graph* (Ctrl-G) from the menu bar. The following dialog will appear.

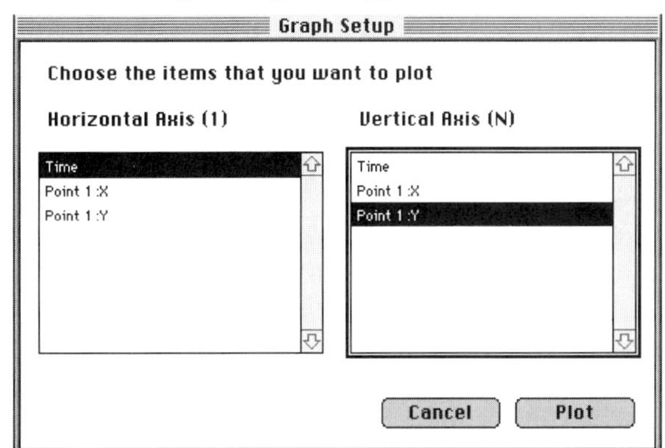

Fig. H.5. The Graph Setup dialog box.

From the left list, choose one value that you want as your horizontal (domain) axis. For this example, choose "Time". You can then choose one or more values in the right list for the vertical (range) axis. For this example, choose "Point s1:*Y*." This will plot *Y* vs. Time for the video point series named, "Point s1." Click "Plot" to create the graph. It should look something like this:

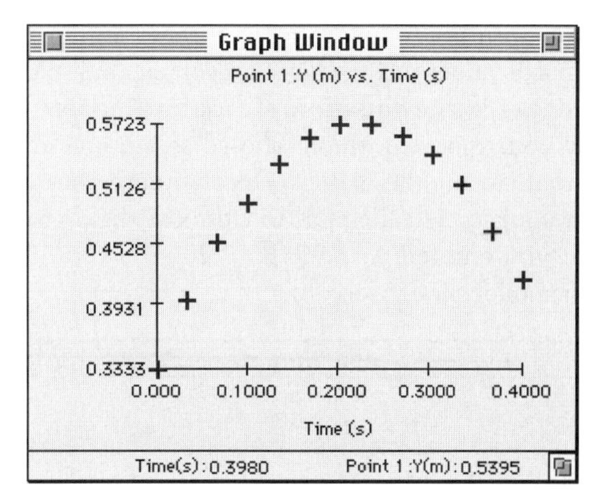

Fig. H.6. A Graph window showing a plot of point 1*Y(m)* vs. Time(s) for a projectile.

Repeat the graph process for an *X* vs. Time plot for Point s1.

CREATING A MATHEMATICAL MODEL OF THE DATA

Since many of the motions of interest in the study of physics can be described by analytic functions, VideoPoint has a graphical modeling feature that enables you to try to develop a mathematical model for a motion. You can do this by comparing a graph of the motion to a graph of an equation. For example, in analyzing a movie of a bungee jumper in free fall, you could select a graph of the experimentally determined values of y vs. t. Then you could choose to model the data with a quadratic equation and then match the parabolic line to the data by changing values of the equation's coefficients. Is the coefficient of the t^2 term close to 4.9 m/s^2?

To model *Y* vs. Time for the projectile in Figure H.6., click on the graph to bring it to the front. Then choose *Graph→Add/Edit Model.* This will bring up the dialog box shown in Fig. H.7.

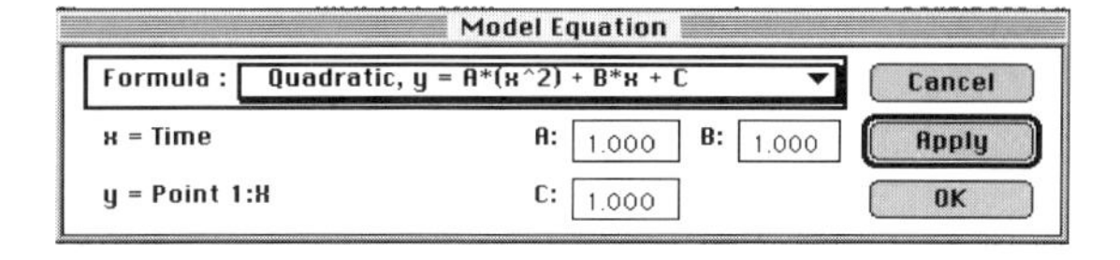

Fig. H.7. The Modeling dialog box.

Since *Y* vs. Time is a parabola for this projectile, choose the "Quadratic" formula. Enter appropriate values in each box that correspond to the constants in the equation. Click on "Apply" to view your modeled graph (shown as a green line) without closing the dialog. Once you are satisfied with your model, click OK to close the dialog box.

If you entered A:-4.90, B:=2.20, and C:=.330, your model should look like that in Fig. H.8.

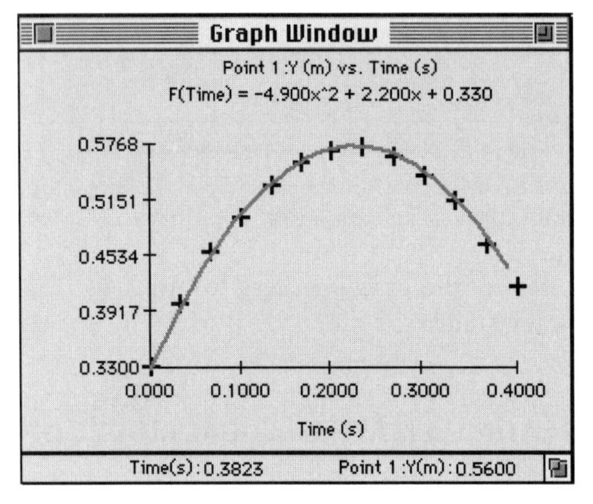

Fig. H.8. The graph model of y vs. time for the ball toss show data points and the line that is a graphical representation of the mathematical model of the data.

The constant A (for the x^2 term) should be approximately –4.9 since this is a projectile that can be modeled by the equation

$y = 1/2g * t\text{^}2 + v_0 t + x_0$.

VIEWING THE DATA IN A TABLE

If you want to view the data that you have taken, click on the table window icon or choose *View→Data Table.* You can select and copy any portion of the data; clicking on the (x) and (y) headers will select entire columns.

TableWindow

Time (s)	Point s1 (pixels) [Origin 1]	
t	x	y
0.000	104.0	106.0
0.03333	146.0	126.0
0.06667	190.0	144.0
0.1000	230.0	160.0
0.1333	272.0	170.0
0.1667	314.0	178.0

Figure H.9. The Table window showing the coordinate data of Point Series I in Cartesian coordinates.

SAVING YOUR WORK

Choose *File→Save.* Decide where you want to save your file.

> What information is saved with a file? The file contains all your data and open windows as well as the name and location of the movie file. It does NOT contain or change the movie file. This keeps the file sizes small and allows files to be associated with movies that are stored on "read-only" networks.

You have successfully analyzed a movie with VideoPoint. Congratulations.

VideoPoint was designed to handle the analysis of more complex two-dimensional motions. You can consult the VideoPoint User's Guide to learn how to analyze movies for which you would like to define more than one video point series or use different coordinate systems. The User's Guide also contains information about how to deal with movies taken with a moving or zooming camera.

APPENDIX I
INTRODUCTION TO MATHEMATICA® SOFTWARE

OVERVIEW

Mathematica is an extensive software package for performing mathematical manipulations and calculations symbolically, graphically, and numerically on a digital computer. Computer algebra systems like Mathematica were developed in the 1950s by scientists interested in artificial intelligence. There are several other computer algebra systems besides Mathematica including Maple, Derive, and Theorist.

Mathematica has over 1500 commands. Among other things it can:

- factor polynomials
- find solutions to polynomial, exponential, trigonometric, and logarithmic equations
- solve systems of linear equations
- graph functions
- differentiate functions
- integrate functions
- take limits
- perform matrix operations

OPENING MATHEMATICA ON THE COMPUTER

To use Mathematica on the computer, you must:

a. Turn on a computer.

b. Use the mouse to open the Mathematica application.

c. Once the Mathematica program is loaded, you will see a blank page. You must enter a command and then press the enter key to execute the command.

For example, suppose you want to take the derivative of the function

$$y = 2x^2 + 3x + 1$$

Then after the • prompt you would enter

```
D[2x^2+3*x+1,x]
```

and then press the <enter> key.

> **Note:** The <enter> key is not the same as the <return> key.

Mathematica will return the symbolic form of the derivative to you in the form $4x + 3$.

> **Warning:** Mathematica is case sensitive. For example, *D* is a recognized command while *d* is not.

A MATHEMATICA SAMPLER

The rundown below is a quick introduction to the software package. For more information you should consult documentation published by Wolfram Research, Inc. On-line documentation includes:

1. The OPEN ME file by using the open file command in file window when Mathematica is open.
2. The Basic Help file by using the open file command in file window when Mathematica is open.
3. The Mathematica Reference Manual.

Numeric Calculations

To use Mathematica as a sophisticated calculator, simply enter the calculation on a single line. Some of the operators available are: + (add), – (subtract), * (multiply), / (divide), ^ (exponentiate), and sqrt() (square root). For example,

```
32*12^4<enter>

In[1]:=32*12^4
Out[1]=663552
```

Mathematica recognizes a number of operators including factorial, greatest common divisor, least common multiple, etc. For example,

```
20!-12!<enter>

In[2]:=20!-12!
Out[2]=2432902007697638400
```

Numeric Values of Common Functions and Constants

Mathematica can calculate the numeric values for functions and constants such as e and π to as many significant figures as you specify:

```
Exp[1.0]<enter>

In[3]:=Exp[1.0]
Out[3]=2.71828

N[Pi,20]<enter>

In[4]:=N[Pi,20]
Out[4]=3.14159265358979323846
```

Solving Equations

You can use Mathematica to solve algebraic equations (or expressions assumed to be equal to zero):

```
e1=a*x^2+b*x+c<enter>

In[5]:= e1=a*x^2+b*x+c
                     2
Out[5]=c + b x + a x

Solve[e1==0,x]<enter>

In[6]:=Solve[e1==0,x]
Out[6]=
                   2
      -b - Sqrt[b  - 4 a c]
{{x -> ---------------------},
               2 a

                 2
      -b + Sqrt[b  - 4 a c]
{x -> ---------------------}}
               2 a
```

Solving Simultaneous Linear Equations

Mathematica is also capable of solving systems of equations. For example, suppose we are trying to use Kirchhoff's laws to solve for the currents in different branches of a complex circuit and we come up with the following three simultaneous linear equations:

1. $I_1 - I_2 = I_3$
2. $20\, I_1 + 100\, I_2 = -4.5$
3. $100\, I_2 - 50\, I_3 = 3.0$

An expression that can be entered into Mathematica to get solutions is given by

```
Solve[{i1-i2==i3, 20*i1+100*i2==-4.5,
100*i2-50*i3==3.0},{i1,i2,i3}]<enter>

In[7]:=Solve[{i1-i2==i3,20*i1+100*i2==-4.5,
100*i2-50*i3==3.0},{i1,i2,i3}]

Out[7]={{i1 -> -0.121875,
         i3 -> -0.10125,
         i2 -> -0.020625}}
```

Algebraic Operations

Mathematica is most powerful when working as a symbolic or algebraic calculator. For example, if an expression is entered, Mathematica will try to simplify it:

```
(x+y)^3*(x+y)^2<enter>

In[8]:=(x+y)^3*(x+y)^2
                  5
Out[8]= (x + y)
```

Calculus

Mathematica is an outstanding tool for studying calculus. You will be able to compute (directional) limits as well as integrate and differentiate:

```
f=(2*x+3)/(7*x+5)<enter>

In[9]:=f=(2*x+3)/(7*x+5)
       3 + 2 x
Out[9]=-------
       5 + 7 x

Limit[f,x->Infinity]<enter>

In[10]:=Limit[f,x->Infinity]
         2
Out[10]=---
         7

f=x*Sin[x]+2*x^2<enter>

In[11]:= f=x*Sin[x]+2*x^2
           2
Out[11]=2 x  + x Sin[x]

D[f,x]<enter>
```

```
In[12]:=D[f,x]
Out[12]=4 x + x Cos[x] + Sin[x]
```

```
Integrate[Sin[x]+x*Cos[x]+4*x,x]<enter>
```

```
In[13]:=Integrate[Sin[x]+x*Cos[x]+4*x,x]
           2
Out[13]=2 x  + x Sin[x]
```

Graphing

The Mathematica plot command provides support for two-dimensional graphs of one or more functions specified as expressions, procedures, parametric functions, or lists of points. In the first plot command, the brackets [ ] indicate a parametric plot:

```
Plot[y^2,{y,-3,3}]<enter>
```

```
In[14]:=Plot[y^2,{y,-3,3}]
```

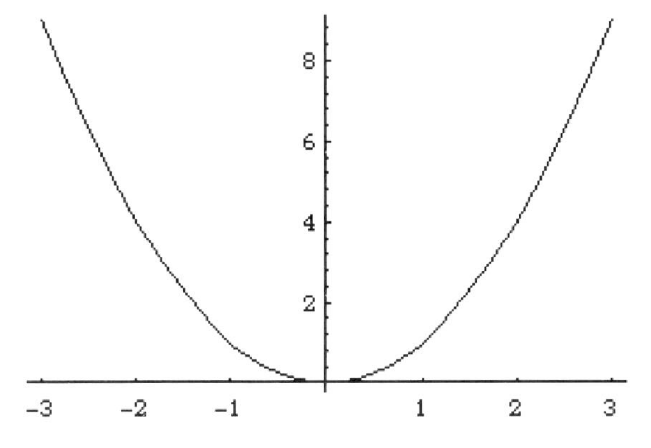


```
Out[14]=-Graphics-
```

The following command uses braces {} to define a set of expressions in x to be plotted:

```
Plot[{4*Sin[x]/(x+1),8*Sin[x]/(x+1)},
{x,0,4*Pi}]<enter>
```

```
In[15]:=Plot[{4*Sin[x]/(x+1),8*Sin[x]/
(x+1)},{x,0,4*Pi}]
```

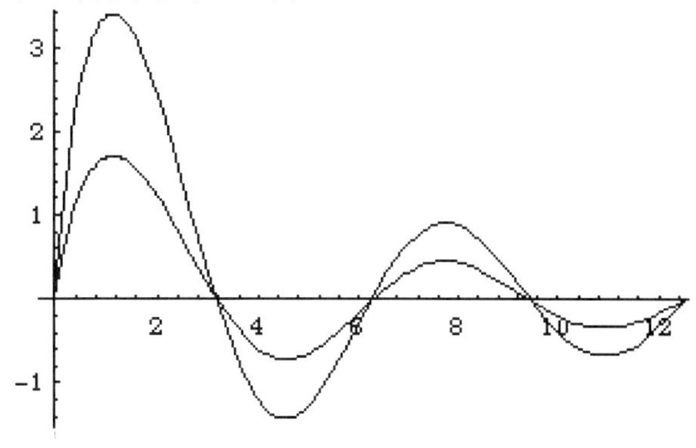


```
Out[15]=-Graphics-
```

Differential Equations

Mathematica can also solve differential equations. A simple example is $y'=y$:

```
DSolve[y'[x]-y[x]==0,y[x],x]<enter>
```

```
In[16]:=DSolve[y'[x]-y[x]==0,y[x],x]
                   x
Out[16]={{y[x] -> E  C[1]}}
```

APPENDIX J
WORKSHOP PHYSICS TABLES

PHYSICAL PROPERTIES

Air (at room temperature and sea level atmospheric pressure)	
Density	1.20 kg/m^3
Specific heat (c_p)	1.00×10^3 J kg^{-1} K^{-1}
Speed of sound	343 m/s
Water (at room temperature and sea level atmospheric pressure)	
Density	1.00×10^3 kg/m^3
Specific heat	4.18×10^3 J kg^{-1} K^{-1}
Speed of sound	1.26×10^3 m/s
Earth	
Density (mean)	5.49×10^3 kg/m^3
Radius (mean)	6.37×10^6 m
Mass	5.97×10^{24} kg
Atmospheric pressure	1.01×10^5 Pa (average sea level)
Mean earth–moon distance	3.84×10^8 m

SI MULTIPLIERS

Abbreviation	Name	Value
G	giga	10^9
M	mega	10^6
k	kilo	10^3
c	cent.	10^{-2}
m	milli	10^{-3}
μ	micro	10^{-6}
n	nano	10^{-9}
p	pico	10^{-12}

PHYSICAL CONSTANTS

Average acceleration of gravity	a_g*	9.80m/s^2
Gravitational constant	G	6.67×10^{-11} N m^{2}kg^{-2}
Electron mass	m_e	9.11×10^{-31} kg
Proton mass	m_p	1.673×10^{-27} kg
Neutron mass	m_n	1.675×10^{-27} kg
Speed of light	c	3.00×10^8 m/s
Universal gas constant	R	8.31 J mol^{-1} K^{-1}
Boltzmann's constant	k	1.38×10^{-23} J/K
Avogadro's number	N_A	6.02×10^{23} mol^{-1}
Permittivity constant	ε_o	8.85×10^{-12} F/m
	$\frac{1}{(4\pi\varepsilon_o)}$	8.99×10^9 N m^{2}C^{-2}
Elementary charge	e	1.60×10^{-19} C
Permeability constant	μ_o	$4\pi \times 10^{-7}$ T m A^{-1}
	μ_o	1.26×10^{-6} N/A^2
Electron volt	eV	1.60×10^{-19} J
Unified atomic mass unit	u	1.66×10^{-27} kg

* This is often denoted g in other texts.

SOLAR SYSTEM

Body	Mean radius of orbit (m)	Mean radius of body (m)	Mass (kg)
Sun		6.96×10^8	1.99×10^{30}
Mercury	5.79×10^{10}	2.42×10^6	3.35×10^{23}
Venus	1.08×10^{11}	6.10×10^6	4.89×10^{24}
Earth	1.50×10^{11}	6.37×10^6	5.97×10^{24}
Mars	2.28×10^{11}	3.38×10^6	6.46×10^{23}
Jupiter	7.78×10^{11}	7.13×10^7	1.90×10^{27}
Saturn	1.43×10^{12}	6.04×10^7	5.69×10^{26}
Moon	3.84×10^8	1.74×10^6	7.35×10^{22}

CONVERSIONS

Length

1 in. = 2.54 cm
1 m = 39.37 in. = 3.281 ft
1 ft. = 0.3048 m
12 in. = 1 ft
3 ft = 1 yd
1 yd = 0.9144 m
1 km = 0.621 mi
1 mi = 1.609 km
1 mi = 5280 ft
1 Å = 10^{-10} m
1 μm = 1 μ = 10^{-6} m = 10^4 Å
1 lightyear = 9.461×10^{15} m

Area

1 m^2 = 10^4 cm^2 = 10.76 ft^2
1 ft^2 = 0.0929 m^2 = 1.44 $in.^2$
1 $in.^2$ = 6.452 cm^2

Volume

1 m^3 = 10^6 cm^3 = 6.102×10^4 $in.^3$
1 ft^3 =1728 $in.^3$ = 2.83×10^{-2} m^3
1 liter = 1000 cm^3 = 1.0576 qt = 0.0353 ft^3
1 ft^3 = 7.481 gal = 28.32 liters = 2.832×10^{-2} m^3
1 gal = 3.786 liters = 231 $in.^3$

Pressure

1 bar = 10^5 N/m^2 = 14.50 $lb/in.^2$
1 atm = 760 mm Hg = 76.0 cm Hg
1 atm = 14.7 lb/in^2 = 1.013 x 10^5 N/m^2
1 Pa = 1 N/m^2 = 1.45 x 10^{-4} lb/in^2

Mass

1000 kg = 1 t (metric ton)
1 slug = 14.59 kg
1 u = 1.66×10^{-27} kg

Force

1 N = 10^5 dyne = 0.2248 lb
1 lb = 4.448 N
1 dyne = 10^{-5} N = 2.248×10^{-6} lb

Velocity

1 mi/h = 1.47 ft/s = 0.447 m/s = 1.61 km/h
1 m/s = 100 cm/s = 3.281 ft/s
1 mi/min = 60 mi/h = 88 ft/s

Acceleration

1 m/s^2 = 100 cm/s^2 = 3.28 ft/s^2
1 ft/s^2 = 0.3048 m/s^2 = 30.48 cm/s^2

Time

1 year = 365 days = 3.16 x 10^7 s
1 day = 24 h = 1.44 x 10^3 min = 8.64 x 10^4 s

Energy

1 J = 0.738 ft·lb = 10^7 ergs
1 cal = 4.186 J
1 Btu = 252 cal = 1.054 x 10^3 J
1 eV = 1.6 x 10^{-19} J
931.5 MeV is equivalent to 1 u
1 kWh = 3.60 x 10^6 J

Power

1 hp = 550 ft·lb/s = 0.746 kW
1 W = 1 J/s = 0.738 ft·lb/s
1 Btu/h = 0.293 W

UNITS

Quantity	Name of unit	In terms of base units	In other common terms
Capacitance (C)	farad (F)	$kg^{-1} \cdot m^{-2} \cdot s^4 \cdot A^2$	C/V
Electric charge (q)	coulomb (C)	$s \cdot A$	
Electric field (**E**)		$kg \cdot m \cdot s^{-3} \cdot A^{-1}$	N/C or V/m
Electric potential (V) (also emf [ε])	volt (V)	$kg \cdot m^2 \cdot s^{-3} \cdot A^{-1}$	J/C or W/A
Electric resistance (R)	ohm (Ω)	$kg \cdot m^2 \cdot s^{-3} \cdot A^{-2}$	V/A
Energy (E)	joule (J)	$kg \cdot m^2/s^2$	N · m
Force (**F**)	newton (N)	$kg \cdot m/s^2$	
Frequency (f)	hertz (Hz)	s^{-1}	
Magnetic field (**B**)	tesla (T)	$kg \cdot s^{-2} \cdot A^{-1}$	Wb/m^2
Magnetic flux (Φ_B)	weber (Wb)	$kg \cdot m^2 \cdot s^{-2} \cdot A^{-1}$	V · s
Power (P)	watt (W)	$kg \cdot m^2/s^3$	J/s
Pressure (p)	pascal (Pa)	$kg \cdot m^{-1} s^{-2}$	N/m^2 or J/m^3

SOME SYMBOLS USED IN THIS ACTIVITY GUIDE

- **a** acceleration
- A area
- a_g acceleration of gravity
- **B** magnetic field
- β electron
- c specific heat (per unit mass)
- C molar heat capacity; capacitance
- e base of natural logs, 2.71828. . . ; magnitude of charge of electron
- E energy
- **E** electric field
- $\mathcal{E}$ emf (electromotive force)
- ε_0 Permittivity of free space
- **F** force
- G gravitational constant
- i current (signed)
- $\hat{x}$ unit vector in x direction
- I rotational inertia; current
- $\hat{y}$ unit vector in y direction
- **J** impulse, current density
- k Boltzmann's constant; spring constant
- K dielectric constant
- $\hat{z}$ unit vector in z direction
- E_{kin} kinetic energy
- L latent heat
- **L** angular momentum of system
- m mass
- M mass
- n number of moles; number of density of charge carriers; number of turns per unit length
- N neutron number of nucleus; number of turns in a coil
- **p** momentum of particle
- P power; pressure
- q charge of particle
- Q heat; charge of system
- $\hat{r}$ position vector
- R universal gas constant; resistance
- S surface area; entropy
- ***S*** surface vector
- t time
- T temperature; period
- E_{pot} potential energy
- v speed
- ***v*** velocity
- V electric potential; potential difference
- W work
- Z atomic number of nucleus
- α magnitude of angular acceleration; alpha particle
- γ ratio of specific heats, c_p/c_v
- λ wavelength; decay constant; charge per unit length
- μ coefficient of friction; mass per unit length; permeability
- f frequency; number of degrees of freedom
- ρ mass (or charge) per unit volume; resistivity
- σ charge per unit area; conductivity
- τ torque magnitude; time constant; mean time between collisions
- Φ flux of a vector field
- ø phase constant
- ω angular speed; angular frequency

VECTOR PRODUCTS

$$\boldsymbol{A} = \vec{A} = A_x\hat{x} + A_y\hat{y} + A_z\hat{z}$$

$$\boldsymbol{B} = \vec{B} = B_x\hat{x} + B_y\hat{y} + B_z\hat{z}$$

$$\vec{A} \bullet \vec{B} = A_x B_x + AB_y + A_zB_z$$

$$\vec{A} \times \vec{B} = \begin{Vmatrix} x & y & z \\ A_x & A_y & A_z \\ B_x & B_y & B_z \end{Vmatrix} = \begin{matrix} (A_yB_z - A_zB_y)\hat{x} \\ -(A_xB_z - A_zB_x)\hat{y} \\ +(A_xB_y - A_yB_x)\hat{z} \end{matrix}$$

THE GREEK ALPHABET

Alpha	Α	α	Eta	Η	η	Nu	Ν	ν	Tau	Τ	τ
Beta	Β	β	Theta	Θ	θ	Xi	Ξ	ξ	Upsilon	Υ	υ
Gamma	Γ	γ	Iota	Ι	ι	Omicron	Ο	ο	Phi	Φ	φ
Delta	Δ	δ	Kappa	Κ	κ	Pi	Π	π	Chi	Χ	χ
Epsilon	Ε	ϵ	Lambda	Λ	λ	Rho	Ρ	ρ	Psi	Ψ	ψ
Zeta	Ζ	ζ	Mu	Μ	μ	Sigma	Σ	σ	Omega	Ω	ω

USEFUL NUMBERS

$\pi = 3.14159$

$e = 2.71828$

$1 \text{ rad} = 57.2958^{\circ}$

$\ln 2 = 0.693147$

$\ln 10 = 2.30259$

$\sin 0 = 0$

$\cos 0 = 1$

$\tan 0 = 0$

$\sin 30^{\circ} = \frac{1}{2}$

$\cos 30^{\circ} = \frac{\sqrt{3}}{2}$

$\tan 30^{\circ} = \frac{1}{\sqrt{3}}$

$(30^{\circ} = \frac{\pi}{6} \text{ rad})$

$\sin 45^{\circ} = \frac{\sqrt{2}}{2}$

$\cos 45^{\circ} = \frac{\sqrt{2}}{2}$

$\tan 45^{\circ} = 1$

$(45^{\circ} = \frac{\pi}{4} \text{ rad})$

$\sin 60^{\circ} = \frac{\sqrt{3}}{2}$

$\cos 60^{\circ} = \frac{1}{2}$

$\tan 60^{\circ} = \sqrt{3}$

$(60^{\circ} = \frac{\pi}{3} \text{ rad})$

$\sin 90^{\circ} = 1$

$\cos 90^{\circ} = 0$

$\tan 90^{\circ} = \infty$

$(90^{\circ} = \frac{\pi}{2} \text{ rad})$

TRIGONOMETRY

Definition of angle (in radians): $\theta = \frac{s}{r}$

2π radians in complete circle

1 radian = 57.3°

TRIGONOMETRIC FUNCTIONS

$\sin\theta = \frac{y}{r}$

$\cos\theta = \frac{x}{r}$

$\tan\theta = \frac{\sin 0}{\cos 0} = \frac{y}{x}$

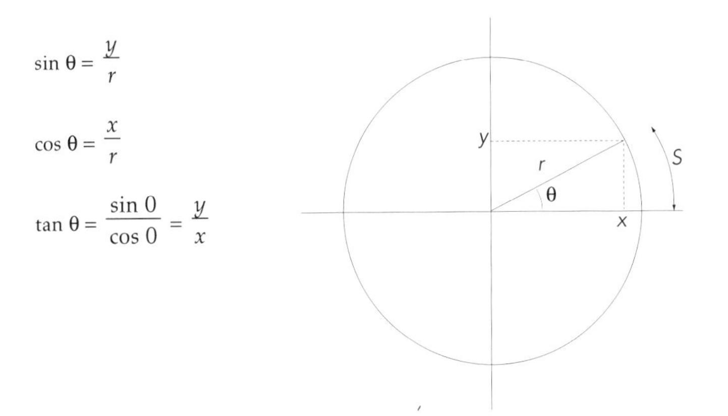

PHOTO CREDITS

Unit 1. Courtesy Priscilla Laws.

Unit 2. Photofest.

Unit 3. Stephen Dalton/Photo Researchers.

Unit 4. Courtesy Priscilla Laws.

Unit 5. Courtesy NASA.

Unit 6. François Gohier/Photo Researchers.

Unit 7. Courtesy Priscilla Laws.

Unit 8. Bill Smith/The Sentinel, Carlisle, Pennsylvania.

Unit 9. Frank Siteman/Stock, Boston.

Unit 10. Tony Savino/The Image Works.

Unit 11. Martha Swope/Life Picture Service.

Unit 12. Martha Swope/Life Picture Service.

Unit 13. Opener: Jean-Claude Lejeune/Stock, Boston.
Figure 13.10: Steve E. Sutton/Duomo Photography, Inc.

Unit 14. Robert Mathena/Fundamental Photographs.

Unit 15. © 1992 Rick Smolan/From *Alice to Ocean.*

Unit 16. Courtesy Priscilla Laws.

Unit 17. Richard Rowan/Photo Researchers.

Unit 18. Courtesy Priscilla Laws.

Unit 19. Courtesy Priscilla Laws.

Unit 20. Courtesy Donna Cox, National Center for Supercomputing Applications, University of Illinois, Urbana-Champaign.

Unit 21. Courtesy Mount Wilson and Palomar Observatories.

Unit 22. Courtesy Priscilla Laws.

Unit 23. Gordon Garradd/Science Photo Library/Photo Researchers.

Unit 24. Courtesy Priscilla Laws.

Unit 25. Courtesy Priscilla Laws.

Unit 26. Opener: Courtesy of International Business Machines Corporation.
Figure 26.19: Simon Bruty/Allsport.

Unit 27. Photo by Dan Jack, courtesy Berkshire Transformer Corporation.

Unit 28. Opener: Courtesy Sargent-Welch, a VWR Company. Inset: Courtesy Lawrence Livermore Laboratory.

INDEX